INTERNATIONAL CENTRE FOR MECHANICAL SCIENCES

COURSES AND LECTURES - No. 218

OCTAV ONICESCU
MEMBER OF THE ROMANIAN
ACADEMY

INVARIANTIVE MECHANICS

SPRINGER-VERLAG WIEN GMBH

Originally published by Springer-Verlag Wien New York in 1975

ISBN 978-3-211-81349-2 ISBN 978-3-7091-2989-0 (eBook)
DOI 10.1007/978-3-7091-2989-0

PREFACE

Mechanics is the science of motion of the bodies of the material universe. For centuries, this frame of our experience has been conceived in various manners.

Sometimes in a very complete and precise manner, in the sense that the universe includes both stars, the solar system and the bodies of the earthly experiments. It was the case of Aristotle's mechanics and also that of Ptolemy's; but each of these three worlds conserved its special motion laws. A proper and universal mechanical principle, unique for the whole universe, was formulated for the first time by Arhimede, who did not try to build a corresponding theory. I am thinking of the principle of the lever that was, however, a universal principle of equilibrium between action and reaction.

In modern times the idea of material identity among all the bodies of the universe made its way; it had a first great victory with the copernican theory. It also has been recorded as an indisputable truth in Leonardo's manuscripts, and it reached the final victory with Galilei's celestial discoveries.

The science of natural motion began as early as people became convinced of a substantial identity among the bodies of our universe. Limited, first, with Galilei, at the motion of the bodies under the strength of their weight here on the Earth, then with Newton for all motions on the Earth and in the planetary system, it has since aspired to include the whole universe. The latter being conceived as a unity realized by the motion of all its material bodies, in a system of interactions which maintain its stability.

The astronomic discoveries which happened in an accelerated rhythm since Newton to our days have strengthened this idea of unity of the universe which reached its culminating point in the discovery of the dilatation phenomenon governed by Hubble's law in its general sense.

To this success of the experimental knowledge, corresponds the creation of the theory of relativity, which gave us a handy geometrical image of this universe in its spatial wholeness as a representation of its material structure and of its general properties.

Coming back to the position of a science of motion in which the presence of the whole universe is to be found in each of its components, not by structural geometric ways, as in relativity, but by means of the analysis of the elementary processes of motion following the line of Newtonian thought, the Invariant

Mechanics, without leaving the spatio- temporal frame of the old science, has found, together with gravity, a second inertial interaction, similar to an elastic repulsive force. Slightly sensitive to current distances, but very sensitive to intergalactic distances, this interaction is for a great part responsible for Hubble's expansion and at the same time for the stability of the universe, in its limits, necessarily finite.

The elaboration of this doctrine began with the studies "A New Mechanics of Material Systems" (Revista Universitatii si Politechnicii din Bucuresti, nr. 3, 1954), "Introduction à une mecanique invariante des systèmes (Revue de Math. pures et appliquées, Bucharest, t.V., 1957), "Une mécanique des systèmes inertiaux. Une théorie de la gravitation. Une mécanique des petites distances" (Journal of Math. and Mech., t.5, no.7, 1958).

These papers have been followed by others dealing with continuous systems as for example "Die Mechanik des starren Körpers", (Revue de mécanique appliquée, Bucharest, t.II, nr. 3, 1958); "La mécanique de certaines particules stables" (Rediconti Sem. Math. Univ. Padova, t.28, 1958), "On The Two Bodies Problem" (Revue de Math. Pures et Appliquées, Bucharest, t.V. nr. 1, 1960), "L'Univers antiminkowskien" (Revue de Physique Acad. R.P.R., Bucharest, t.V., nr. 3–4, 1961), as the Lectures at the Institute for Mechanics – University of Triest – 1966, when I gave the name of "Invariantive Mechanics" to this new theory.

A synthesis of these theories was given in the volume "Mecanica invariantiva si Cosmologia" Editura Acadamiei R.S.R., 1974. On this occasion was established that the theoretical law of conservation of the impulse of the whole system of bodies of the Universe corresponds to Hubble's empirical law (C.R. Acad. Sci. Paris 1972). In the same volume was incorporated Mihaila's correction of the gravity law.

The full content of the present book was delivered as lectures at the "International Center of Mechanical Sciences" Udine, Italy, to whose leadership I am honored to appartain.

I express my friendly thanks to the staff and the personnel of the publishing office of the Center for the accuracy shown by the preparation of the favorable issue of my work.

Octav Onicescu

INTRODUCTION

The principles of Newton's Natural Philosophy have a vast field of application which includes the motion of bodies at velocities extending from zero to the velocities of planetary motion, and involving distances which approach the large scale and the dimensions of our Galaxy and reach the threshold of the nuclear universe at the opposite end. Only second order effects of planetary motion are outside the realm of Newton's Natural Philosophy, which is, within bounds to be ascertained, a science of nature.

A science of motion, where velocities and distances may exceed the limits assigned above, requires a new structure which should be reduced to Newtonian one for velocities and distances consistent with it; it is only beyond these limits that quantitative differences should become apparent and significant thus allowing the new structure to be still considered, in a broader sense, a science of nature.

In order to build this new Mechanics it is necessary to reconsider the simplest problems of motion, of inertial motion first, of a single material particle, then of two or more material particles, for systems such as the solid body of the Newtonian mechanics and ultimately of the motion in a field.

The inertial motion of a material particle reveals an unexpectedly vast content of the inertia of a material mass in motion including its Einsteinean characteristics and other features besides. The law of inertial motion of a system of material particles appears as a theory of gravitation and at the same time of the expansion of the universe.

Likewise, the introduction of the field discloses a wealth of possible forms which are at the disposal of the physicist and require only to be recognized and eventually classified according to specific features which elude, at least on a first examination, the criteria of mechanics.

Throughout this investigation the representative space of mechanics is Newton's four-dimensional space $S_N = E_3 \times T$, i.e. the space consisting of the Euclidean space isotropic, homogeneous with three dimensions associated with the Newtonian time, homogeneous and uniform, which is itself a one-dimensional length and angles (in the case of E_3).

The principles of the invariant mechanics emerge and are stated as they are required by increasingly complex mechanical systems.

The simplest material system in Newtonian mechanics is the material point.

Hence, the inertial motion of a material point must be studied first, and, in connection with it, the motion under the effect of a field.

We shall next study all the aspects of the inertial motion of a system consisting of two material masses.

We shall give but a short account on the motion in a field of a system of two material particles; it will be shown that in general we have to deal with interactions of a Coulombian type the form of which will be given.

One paragraph is devoted to the motion, both inertial and in a field, of a body of the solid particle type.

The last paragraph deals with the motion of the perihelion and is due to Dr. I. Mihaila.

CHAPTER I

MOTION OF A MATERIAL POINT

§ 1. Inertial Motion

1. Material Point. The material point as an object of mechanics is characterized by two vectors of the four-dimensional space S_N:

a. The position-time vector $(\mathbf{x}, t)$, $\mathbf{x}$ being the vector with the components x_1, x_2, x_3 in the 3-dimensional Euclidean space E_3. The reference systems in the space E_3 and in time are the Newtonian inertial reference systems.

b. The impulse-energy vector $(\mathbf{p}, E)$, where the impulse has the components p_1, p_2, p_3 and $E = E\ (\alpha)$ is a function, unknown yet, of the invariant

$$\alpha = 1/2\,(p_1^2 + p_2^2 + p_3^2) = 1/2\,p^2 \ . \tag{1}$$

2. The motion of the material point characterized by these elements is indicated by the differential operator d which is defined with the help of the geometrical differential of the impulse-energy vector.

The geometrical differential of the vector $(\mathbf{p}, E)$ is defined with the help of the external derivative of the form

$$\omega_\delta = \mathbf{p}.\delta\mathbf{x} - E\delta t\,, \tag{2}$$

namely

$$D\omega_\delta = d\omega_\delta - \delta\omega_d = d\mathbf{p}.\delta\mathbf{x} - dE.\delta t - d\mathbf{x}\delta\mathbf{p} + dt\delta E\,,$$

i.e.

$$D\omega_\delta = d\mathbf{p}\,\delta\mathbf{x} + (E'(\alpha)\,\mathbf{p}\,dt - d\mathbf{x})\,\delta\mathbf{p} - dE\,\delta t \ . \tag{3}$$

The components of the geometrical differential $D(\mathbf{p},E)$ of the impulse- energy vector are represented by the coefficients of the variations $\delta\mathbf{x}$, $\delta\mathbf{p}$, δt in the expression of $D\omega_\delta$, namely

(4) $$dp,\ E'(\alpha)\mathbf{p}\,dt - d\mathbf{x},\ dE\ .$$

3. Inertial motion. Principle 1. The motion operator d is defined by the condition that the geometrical differential of the impulse-energy vector

(5) $$D(\mathbf{p}, E) = 0\ .$$

This gives the equations

(6) $$\frac{d\mathbf{p}}{dt} = 0,\quad E'(\alpha)\mathbf{p} = \frac{d\mathbf{x}}{dt},\quad \frac{dE}{dt} = 0\ .$$

Since the last equation is identical to

$$E'(\alpha)\mathbf{p}\,\frac{d\mathbf{p}}{dt} = 0,$$

it follows that it will be satisfied together with the first equation $d\mathbf{p}/dt = 0$, hence

(7) $$\mathbf{p} = \mathbf{p}_0,\quad E(\alpha) = E(\alpha_0)$$

which means that during the inertial motion the impulse is constant. The first principle may be stated also in the following form:

The motion operator d is defined by the condition that the external derivative $D\omega_\delta = d\omega_\delta - \delta\omega_d$ of the fundamental form ω_δ should vanish for any δ.

The equivalence of the two definitions of the first principle is obvious.

4. The expressions of impulse, mass and energy. The second vector equation of (6) gives for the impulse the expression

(8) $$\mathbf{p} = m\mathbf{v}$$

after putting $\mathbf{v} = \dot{\mathbf{x}}$ and $m = 1/E'(\alpha)$, where $E'(\alpha) \neq 0$, a condition which will be verified subsequently.

In expression (8) m represents the mass of the material point.

Using (8), the fundamental form ω_δ may be written

$$\omega_\delta = m(\mathbf{v}.\delta\mathbf{x} - \frac{E}{m}\,\delta t) \ . \tag{9}$$

We now remark that on the one hand from the statement of the law of motion, it follows that the form ω_δ includes all the elements of motion, and on the other hand that, since the space S_N is uniform and isotropic, the motion of the material point, when it passes from one position to another, differs, besides position and velocity, only by mass. This means that the expression $\mathbf{v}.\delta\mathbf{x} - (E/m)\delta t$ of (9) is the same for any material particle; this remark will be stated in the following

Postulate. The ratio E/m is a universal constant.

Since this ratio has the homogeneity of the square of a velocity we put

$$\frac{E}{m} = \epsilon\omega^2 \tag{10}$$

with $\epsilon = \pm 1$. Hence

$$E = \epsilon m \omega^2 \ . \tag{11}$$

5. Dependency of mass on velocity. By the second form of the principle which gives the law of motion, we have

$$D\omega_\delta = 0$$

or, taking into account (3) and (7), it remains

$$\mathbf{x}.\,\delta\mathbf{p} = E'(\alpha)\,\mathbf{p}.\,\delta\mathbf{p} = E'(\alpha)\,\delta\alpha = \delta\,E(\alpha)$$

or, using (8) and (11),

$$\mathbf{p}.\,\delta\mathbf{p} = \epsilon\omega^2\, m\,\delta m \ .$$

From this we derive $\delta(\mathbf{p}^2 - \epsilon\omega^2 m^2) = 0$ and consequently

$$\mathbf{p}^2 - \epsilon\omega^2 m^2 = -\epsilon\omega^2 m_0^2 \, , \tag{12}$$

where m_0 is the mass corresponding to the value $p_0 = 0$ of the impulse and hence, – by (8) – to the velocity $v_0 = 0$; therefore m_0 is the mass at rest.

From (12) we obtain first the expression

$$m = \sqrt{m_0^2 + \frac{\epsilon}{\omega^2} p^2} \; . \tag{13}$$

Replacing in (12) p by mv , we obtain the expression of the mass as a function of velocity

$$m = \frac{m_0}{\sqrt{1 - \epsilon \frac{v^2}{\omega^2}}} \; . \tag{14}$$

Hence we have to examine two cases: 1°. $\epsilon = +1$. In this case relation (14) becomes

$$m = \frac{m_0}{\sqrt{1 - \frac{v^2}{\omega^2}}} \tag{15}$$

and ω appears as a limiting velocity. Experimental evidence shows that this limiting velocity is c, the velocity of light in vacuum. Hence we are led to put $\omega = c$ in order to comply with experimental evidence, thus obtaining the expression

$$m = \frac{m_0}{\sqrt{1 - \frac{v^2}{c^2}}} \; , \tag{16}$$

which corresponds to the expression of mass as a function of velocity, a relation established by Lorentz and found by Einstein in a different way.

With this value of ϵ and the respective necessary value of ω, formula (13) becomes

$$m = \sqrt{m_0^2 + \frac{1}{c^2} p^2} \; . \tag{17}$$

In this case, using (11) and (1), we obtain

$$E = E(\alpha) = mc^2 = c^2\sqrt{m_0^2 + \frac{1}{c^2} p^2} . \qquad (18)$$

The energy is therefore expressed by mc^2 and at the same time we verify that

$$E'(\alpha) = \frac{1}{\sqrt{m_0^2 + \frac{1}{c^2} p^2}} \neq 0.$$

The choice $\epsilon = 1$ and $\omega = c$ is imposed by experience and emphasizes a structural characteristic of matter in motion through space, hence a relation between matter, space and time.

Thus, the theory presented here goes deep into the mechanical phenomenality; it shows that the inertia of matter is intimately connected to the structure of space and time, which are generally considered only under a geometrical aspect.

The law of inertial motion.

Since, by (15) and (17), we have $\mathbf{p} = m\mathbf{v}$, it follows that $m \neq 0$ if $m_0 \neq 0$ and equations (7) shows that $\mathbf{v}$ is constant during the motion. Hence the motion is rectilinear and uniform. The energy of the material particle also is constant and has the value

$$mc^2 = m_0c^2 + 1/2\, m_0 v^2 + \theta/(1 - v^2/c^2), \quad \text{where } 0 \leqslant |\theta| \leqslant 1,$$

hence the predominant term is the internal energy m_0c^2 associated with the matter represented by the mass at rest m_0.

§2. The Universe of Minkowski-Einstein

1. The Metrics. The form obtained for ω_δ, following the choice of $\epsilon = 1$ and $\omega = c$, becomes

$$(19) \qquad \omega_\delta^{(i)} = m(\mathbf{v}.\delta\mathbf{x} - c^2\,\delta t) = m\,\frac{\mathbf{dx}\;\delta\mathbf{x} - c^2\,dt\,\delta t}{dt}\ .$$

The impulse-energy vector is $(m\mathbf{v},\ mc^2)$, where

$$m = m_0/\sqrt{1 - v^2/c^2}\ .$$

Letting the mass to be the factor, the inertial fundamental form may be written

$$(20) \qquad \omega_\delta^{(i)} = \frac{m\varphi}{dt},$$

thus emphasizing the bilinear form

$$(21) \qquad \varphi = \mathbf{dx}.\delta\mathbf{x} - c^2\,dt\,\delta t$$

which, together with the quadratic form

$$(22) \qquad d\sigma^2 = c^2\,dt^2 - dx_1^2 - dx_2^2 - dx_3^2\ ,$$

is invariant under the Lorentz-Poincaré continuous group of linear transformations with 10 parameters.

We observe that

$$m = \frac{m_0\,dt}{\sqrt{c^2\,dt^2 - \mathbf{dx}^2}} = m_0\,\frac{dt}{d\sigma}$$

hence

$$(23) \qquad m/dt = m_0/d\sigma\ .$$

Therefore, besides the invariance of the form $d\sigma^2$, hence also of the respective form $c^2 dt\delta t - dx.\delta x$, relation (20), which may also be written

$$\omega_\delta^{(i)} = -m_0 \frac{c^2 dt\delta t - dx\,\delta x}{d\sigma}, \tag{24}$$

shows that the form $\omega_\delta^{(i)}$ too is invariant under the Lorentz-Poincaré group.

The expression $c^2 dt\delta t - dx.\delta x$ is a quasi-scalar product for the quasi-metrics

$$c^2 dt^2 - dx_1^2 - dx_2^2 - dx_3^2 .$$

If the four-dimensional space $S[t, x_1, x_2, x_3]$ is endowed with the above metrics, then it is of the Minkowski-Einstein type. The motions (dx_1, dx_2, dx_3, dt) starting from the point $P(t, x_1, x_2, x_3)$ correspond to a real motion if

$$c^2 dt^2 - dx_1^2 - dx_2^2 - dx_3^2 > 0 . \tag{25}$$

The motions for which

$$c^2 dt^2 - dx_1^2 - dx_2^2 - dx_3^2 = 0, \tag{26}$$

hence with the velocity of light, are not implicitly realized by material points; for material points they are limiting motions.

The variety

$$c^2 \xi_0^2 - \xi_1^2 - \xi_2^2 - \xi_3^2 = 0, \tag{27}$$

where $\xi_0, \xi_1, \xi_2, \xi_3$ are the parameters of the direction of motion – for which therefore $dt/\xi_0 = dx_1/\xi_1 = dx_2/\xi_2 = dx_3/\xi_3$ –, is a hypercone with the vertex at the point $P(t, x_1, x_2, x_3)$ in the four-dimensional space $\Sigma(\xi_0, \xi_1, \xi_2, \xi_3)$; it separates this space into two domains; the inner domain is the universe of the motions and is characterized by the inequality

$$c^2 \xi_0^2 - \xi_1^2 - \xi_2^2 - \xi_3^2 > 0 . \tag{28}$$

The moving point $P(t, x_1, x_2, x_3)$ entails with it the separation cone of the universe of motions.

2. The Lorentz Transformations. The linear transformations which leave invariant the form $c^2\xi_0^2 - \xi_1^2$, and for which $\xi_2 = \xi_2^0$, $\xi_3 = \xi_3^0$, have been detected and turned to account by Lorentz in relation of the electromagnetic phenomenology. They are expressed by the relations

$$\begin{aligned} \xi_0' &= \alpha\xi_0 + \beta\xi_1 \\ \xi_1' &= \gamma\xi_0 + \delta\xi_1 \,. \end{aligned} \tag{29}$$

The coefficients of this transformation verify the identity

$$c^2\xi_0'^2 - \xi_1'^2 = c^2\xi_0^2 - \xi_1^2 \,. \tag{30}$$

Then we have an argument φ for which

$$\alpha = \mathrm{ch}\varphi,\ \beta = -\tfrac{1}{c}\mathrm{sh}\varphi$$

$$\gamma = -\mathrm{ch}\varphi,\ \delta = \mathrm{ch}\varphi \,.$$

Putting

$$u = -c\,\frac{\mathrm{sh}\varphi}{\mathrm{ch}\varphi},$$

we obtain

$$\begin{aligned} \xi_0' &= \frac{1}{\sqrt{1-\frac{u^2}{c^2}}}\left(\xi_0 + \frac{u}{c^2}\,\xi_1\right) \\ \xi_1' &= \frac{1}{\sqrt{1-\frac{u^2}{c^2}}}\left(\xi_1 + u\xi_0\right), \end{aligned} \tag{31}$$

whence

$$\frac{\xi'_1}{\xi'_0} = \frac{\dfrac{\xi_1}{\xi_0} + u}{1 + \dfrac{u}{c^2}\dfrac{\xi_1}{\xi_0}} \, . \tag{32}$$

Since ξ_0 and ξ_1 are proportional to dt and dx_1, and putting

$$\frac{\xi_1}{\xi_0} = \frac{dx_1}{dt} = v, \quad \frac{\xi'_1}{\xi'_0} = \frac{dx'_1}{dt'} = v'$$

and interpreting u as a velocity, we obtain from (32) Einstein's law of the composition of velocities

$$v' = \frac{v + u}{1 + \dfrac{vu}{c^2}} \, . \tag{33}$$

This law shows that if $v \leqslant c$, $u \leqslant c$, then we have also for the resultant velocity v' the inequality

$$v' \leqslant c \, .$$

The equality is valid only if one of the velocities v or u is equal to c.

Indeed, from (33) we have

$$\frac{v}{c} + \frac{u}{c} < 1 + \frac{v}{c}\,\frac{u}{c}$$

since this inequality is equivalent to the following one

$$\frac{v}{c}\left(1 - \frac{u}{c}\right) < 1 - \frac{u}{c} \, ,$$

under the assumption $v < c$, $u < c$. If $v = c$, or $u = c$, we have equality.

§3. Waves Associated to the Motion

1. Wave Associated with the Photon Motion. If the mass at rest m_0 of a material point vanishes, we are out of the limiting conditions of the preceding theory and the motion becomes meaningless if it is interpreted exactly within the previous frame. However, it retains a meaning in a wave motion interpretation by the aid of Jacobi's equation, whose complete integral leads to the law of motion.

The object of the motion which attains the velocity of light may retain yet a finite dynamical mass, m, if it has a mass at rest $m_0 = 0$.

Indeed, from the first two of the fundamental relations

$$E = mc^2 , \quad \mathbf{p} = m\mathbf{v}, \quad m = \sqrt{m_0^2 + 1/c^2 \ \mathbf{p}^2}$$

we obtain

$$\mathbf{p} = E \frac{\mathbf{v}}{c^2} ,$$

whence, assuming $v = c$,

$$p = \frac{E}{c} .$$

From the last fundamental relation we obtain, when $m_0 = 0$,

$$m = \frac{p}{c} .$$

Hence, the energy is expressed by

$$E = pc = c\sqrt{p_1^2 + p_2^2 + p_3^2} ,$$

and Jacobi's theorem gives the equation

$$c \sqrt{\left(\frac{\partial\psi}{\partial x_1}\right)^2 + \left(\frac{\partial\psi}{\partial x_2}\right)^2 + \left(\frac{\partial\psi}{\partial x_3}\right)^2} = \frac{\partial\psi}{\partial t}$$

i.e.

$$\left(\frac{\partial\psi}{\partial x_1}\right)^2 + \left(\frac{\partial\psi}{\partial x_2}\right)^2 + \left(\frac{\partial\psi}{\partial x_3}\right)^2 - \frac{1}{c^2}\left(\frac{\partial\psi}{\partial t}\right)^2 = 0 . \tag{34}$$

This is the equation of the characteristics of the partial differential equation

$$\frac{\partial^2 \phi}{\partial x_1^2} + \frac{\partial^2 \phi}{\partial x_2^2} + \frac{\partial^2 \phi}{\partial x_3^2} - \frac{1}{c^2}\frac{\partial^2 \phi}{\partial t^2} = 0, \qquad (35)$$

whose elementary waves are propagated with the velocity of light.

Let us consider the plane waves of equation (35).The phase of such a wave of wave length λ and frequency ν is given by the expression

$$q = \frac{\alpha_1}{\lambda} x_1 + \frac{\alpha_2}{\lambda} x_2 + \frac{\alpha_3}{\lambda} x_3 - \nu t$$

which must be invariant under the same Lorentz-Poincaré group of invariants of equation (35) and of the motion, and hence of the fundamental form

$$\omega_\delta = m\dot{x}_1\,\delta x_1 + m\dot{x}_2\,\delta x_2 + m\dot{x}_3\,\delta x_3 - E\delta t\ .$$

Therefore the two forms q and ω_δ must be identical, except a factor h, a universal constant:

$$\frac{m\dot{x}_1}{\alpha_1/\lambda} = \frac{m\dot{x}_2}{\alpha_2/\lambda} = \frac{m\dot{x}_3}{\alpha_3/\lambda} = \frac{E}{\nu} = h\ .$$

The last equality of the above system may be written

$$E = h\nu$$

and represents the quantic expression of the energy, a multiple of Planck's constant.

2. Wave associated with a moving material point. In this case $m_0 \neq 0$ and by equation (17, § 1) the energy is

$$E = mc^2 = c^2\sqrt{m_0^2 + 1/c^2\,(p_1^2 + p_2^2 + p_3^2)} = c\sqrt{m_0^2\,c^2 + p_1^2 + p_2^2 + p_3^2}\ .$$

Jacobi's equation becomes in this case

$$\sqrt{m_0^2 c^2 + \left(\frac{\partial \psi}{\partial x_1}\right)^2 + \left(\frac{\partial \psi}{\partial x_2}\right)^2 + \left(\frac{\partial \psi}{\partial x_3}\right)^2} = \frac{1}{c} \frac{\partial \psi}{\partial t}$$

or

$$m_0^2 c^2 + \left(\frac{\partial \psi}{\partial x_1}\right)^2 + \left(\frac{\partial \psi}{\partial x_2}\right)^2 + \left(\frac{\partial \psi}{\partial x_3}\right)^2 - \frac{1}{c^2}\left(\frac{\partial \psi}{\partial t}\right)^2 = 0 . \tag{36}$$

Let us make this equation homogeneous with the help of a function $\chi(\psi, t; x_1, x_2, x_3)$ which must vanish when ψ verifies (36). The function χ satisfies, then, the equation

$$\left(\frac{\partial \chi}{\partial x_1}\right)^2 + \left(\frac{\partial \chi}{\partial x_2}\right)^2 + \left(\frac{\partial \chi}{\partial x_3}\right)^2 + m_0^2 c^2 \left(\frac{\partial \chi}{\partial \psi}\right)^2 - \frac{1}{c^2}\left(\frac{\partial \chi}{\partial t}\right)^2 = 0 . \tag{37}$$

We seek for χ a stationary expression. In that case we have

$$\frac{\partial \chi}{\partial t} = E \frac{\partial \chi}{\partial \psi}$$

and hence equation (37) becomes

$$\left(\frac{\partial \chi}{\partial x_1}\right)^2 + \left(\frac{\partial \chi}{\partial x_2}\right)^2 + \left(\frac{\partial \chi}{\partial x_3}\right)^2 - \frac{v^2}{c^4}\left(\frac{\partial \chi}{\partial t}\right)^2 = 0 .$$

The respective wave equation is

$$\Delta\phi - \frac{v^2}{c^4}\frac{\partial^2 \phi}{\partial t^2} = 0 .$$

The velocity of the corresponding wave is

$$w = \frac{c^2}{v}$$

and represents the de Broglie relation between the two kinds of velocities: the corpuscular, v, and the wave velocity w .

The other relations of de Broglie are obtained by identifying – as in the case of photons – the two invariants

$$q = \frac{\alpha_1}{\lambda}\,\delta x_1 + \frac{\alpha_2}{\lambda}\,\delta x_2 + \frac{\alpha_3}{\lambda}\,\delta x_3 - \nu\delta t$$

and

$$\omega_\delta = p_1\,\delta x_1' + p_2\,\delta x_2' + p_3\,x_3' - E\,\delta t,$$

where

$$\delta x_1' = \frac{c^2}{v}\,\delta x_1,\ \delta x_2' = \frac{c^2}{v}\,\delta x_2,\ \delta x_3' = \frac{c^2}{v}\,\delta x_3\ .$$

Hence we obtain the equalities

$$\frac{\lambda p_1}{\alpha_1} = \frac{\lambda p_2}{\alpha_2} = \frac{\lambda p_3}{\alpha_3} = \frac{E}{\nu} = h'\ .$$

Since the length of the two vectors (p_1, p_2, p_3, E) and $(\alpha_1/\lambda, \alpha_2/\lambda, \alpha_3/\lambda, \nu)$ is zero, the ratio h′ does not depend on velocity and remains unaltered when the velocity is equal to the velocity of light; hence we have $h' = h$.

Therefore we have also for the material point

$$E = h\nu, \quad \mathbf{p} = h\mathbf{I},$$

where the vectorial constant **I** may be called the <u>vectorial frequency of propagation.</u>

§4. Motion in a Field

1. Principles. Potential Form and Potential Vector. At this stage of the construction of the theory, there appears an important difference between the Invariantive and Newton's mechanics. Whereas in Newton's mechanics the field is represented by a scalar potential, the invariantive mechanics requires a vector potential in correspondence with the momentum-energy quadrivector of the inertial motion.

Hence, the results obtained previously for inertial motion require not only a formal structure for the law of motion in the presence of a field, but also a structure for the field itself.

Let $A_1(x_1, x_2, x_3, t)$, $A_2(x_1, x_2, x_3, t)$, $A_3(x_1, x_2, x_3, t)$, $A_0(x_0, x_1, x_2, t)$ be a four-dimensional vector potential function of the position $\mathbf{x}(x_1, x_2, x_3)$ and of the time t in the universe $S_N = E_3 \times T$ and the potential form

$$\omega_\delta^{(p)} = A_1 \delta x_1 + A_2 \delta x_2 + A_3 \delta x_3 - A_0 \delta t$$

which is a scalar product, hence an invariant under the transformations which leaves invariant the inertial form established above

$$\omega_\delta^{(i)} = m\dot{x}_1 \delta x_1 + m\dot{x}_2 \delta x_2 + m\dot{x}_3 \delta x_3 - mc^2 \delta t \ .$$

The law of motion is defined by the following principle:

Principle 2. The Cartan derivative of the impulse-energy vector is equal to the Cartan derivative of the potential vector.

By definition the components of the Cartan derivative of the potential vector $(\mathbf{A}, A_0)$ are the coefficients of the external derivative of the potential form which may be written

$$\omega_\delta^{(p)} = \mathbf{A}.\delta\mathbf{x} - A_0 \delta t$$

where $\mathbf{A}$ is the space vector whose components are A_1, A_2, A_3. On this basis we may state the second principle in a form approaching that of d'Alembert in classical mechanics:

The motion represented by the differential operator d is defined by the equality of the external derivatives of the inertial form and of the potential form

$$D\omega_\delta^{(i)} = D\omega_\delta^{(p)}$$

for an arbitrary δ.

2. Characteristic Magnitudes and Relations of the Field. Being geometrical operation, the external derivative of $\omega_\delta{}^{(p)}$ supplies elements of a vectorial character which are invariant under the Lorentz group.

As it can be easily verified, we have

$$D\omega_\delta^{(p)} = d\omega_\delta^{(p)} - \delta\omega_d^{(p)} = E_1(dx_1\delta t - dt\,\delta x_1) + E_2(dx_2\delta t - dt\,\delta x_2) +$$

$$+ E_3(dx_3\delta t - dt\,\delta x_3) + H_1(-dx_3\delta x_2 + dx_2\delta x_3) + H_2(-dx_1\delta x_3 + dx_3\delta x_1) +$$

$$+ H_3(-dx_2\delta x_1 + dx_1\delta x_2),$$

where the vectors $\mathbf{E}(E_1, E_2, E_3)$ and $\mathbf{H}(H_1, H_2, H_3)$ are defined as follows:

$$\mathbf{E} = \text{grad}\, A_0 - \frac{\partial \mathbf{A}}{\partial t} \;\; ; \;\; \mathbf{H} = \text{rot}\, \mathbf{A} \; .$$

From the above relations we obtain the first group of the characteristic field relations

$$\text{rot}\, \mathbf{E} + \frac{\partial \mathbf{H}}{\partial t} = 0,$$

$$\text{div}\, \mathbf{H} = 0 \; .$$

Putting

$$a = \frac{\partial A_1}{\partial x_1} + \frac{\partial A_2}{\partial x_2} + \frac{\partial A_3}{\partial x_3} - \frac{1}{c^2}\frac{\partial A_0}{\partial t} \tag{38}$$

and considering the four-component vector $\Box\mathbf{A}$,

$$\Box A_j = \Delta A_j - \frac{1}{c^2}\frac{\partial^2 A_j}{\partial t^2} \;\; ; \;\; j = 1,2,3,0,$$

it follows that grad a $-\Box\mathbf{A}$ is a four-component vector of the Minkowski space.

The three spatial components are

(39) $$\text{rot } \mathbf{H} + \frac{1}{c^2}\frac{\partial \mathbf{E}}{\partial t} = \text{grad a} - \Box\mathbf{A}$$

and the time component is

(40) $$\text{div } \mathbf{E} = \frac{\partial a}{\delta t} + \Box A_0 \ .$$

The equations of the stationary field are obtained by putting

$$a = 0$$

and

$$\Box\mathbf{A} = 0$$

after replacing the three-dimensional vector $\mathbf{A}$ by the vector $e/c\,\mathbf{A}$,where e represents the elementary charge,and eA_0 replaces A_0.

From (39) and (40) we obtain the second group of field equations

(41) $$\text{rot } \mathbf{H} + \frac{1}{c^2}\frac{\partial \mathbf{E}}{\partial t} = 0,$$

(42) $$\text{div } \mathbf{E} = 0 \ .$$

The potentials $\mathbf{A}$ and A_0 which satisfy the equations (38) of wave propagation constitute the basis of the Maxwell stationary field if we identify $\mathbf{E}$ to an electric field and $\mathbf{H}$ to a magnetic field.

In the case of non-stationary fields, the conditions

(43) $$a = 0, \ \Box\mathbf{A} = -4\pi\mathbf{k}, \ \Box A_0 = -4\pi\theta,$$

where $\mathbf{k}$ is a field vector and θ a scalar, lead to the equations

(44) $$\text{rot } \mathbf{H} + \frac{1}{c^2}\frac{\partial \mathbf{E}}{\partial t} = 4\pi\mathbf{k},$$

$$\operatorname{div} \mathbf{E} = 4\pi\theta \, . \tag{45}$$

3. Equation of Motion. The equation of motion results by applying the second principle.

Since

$$D\omega_\delta^{(i)} = d\mathbf{p}.\delta\mathbf{x} - d(mc^2)\,\delta t$$

and

$$D\omega_\delta^{(p)} = -(\mathbf{E}dt + d\mathbf{x} \times \mathbf{H})\delta\mathbf{x} + \mathbf{E}d\mathbf{x}.\delta t\,,$$

by applying the second principle, we obtain the vectorial equation of motion

$$\frac{d\mathbf{p}}{dt} = -(\mathbf{E} + \mathbf{v} \times \mathbf{H}), \tag{46}$$

and from the energy equation,obtained by equating the coefficients of δt in the two members, there follows

$$\frac{d(mc^2)}{dt} = -\mathbf{E}.\mathbf{v} = -(\mathbf{E} + \mathbf{v} \times \mathbf{H}).\mathbf{v} = \mathbf{v}.\frac{d\mathbf{p}}{dt}, \tag{47}$$

which is satisfied identically if $\mathbf{p} = m\mathbf{v}$ and $m = m_0/\sqrt{1-v^2/c^2}$.

4. Jacobi's Equation of Motion. By putting

$$\pi_j = p_j - \frac{e}{c} A_j \ , \ j = 1,2,3 \tag{48}$$

and

$$E = c\sqrt{m_0^2 c^2 + \pi_1^2 + \pi_2^2 + \pi_3^2} + e\,W \tag{49}$$

we obtain the same conditions that were required for applying the Hamilton-Jacobi method. Substituting in E the values of π_1, π_2, π_3 given by (48), we obtain from (49)

$$E = c\sqrt{m_0^2c^2+\sum_j(p_j - \frac{e}{c}A_j)^2} + eW.$$

Jacobi's equation becomes then,

$$c\sqrt{m_0^2c^2+\sum(\frac{\partial\varphi}{\partial x_j} - \frac{e}{c}A_j)^2} + eW = -\frac{\partial\varphi}{\partial t} .$$

We make it homogeneous by introducing a function $\chi(\varphi, t, x_1, x_2, x_3) = 0$ which gives

$$c\frac{\partial\varphi}{\partial t} = -\frac{\partial\chi}{\partial t}\Big/\frac{\partial\chi}{\partial\varphi} \; ; \; \frac{\partial\varphi}{\partial x_j} = -\frac{\partial\chi}{\partial x_j}\Big/\frac{\partial\chi}{\partial\varphi} \; ; \; j = 1,2,3,$$

whence the equation

$$\sum_j\left(\frac{\partial\chi}{\partial x_j}\right)^2 - \frac{1}{c^2}\left(\frac{\partial\chi}{\partial t}\right)^2 + \frac{e^2}{c^2}\left(\sum A_j^2 - \frac{W^2}{c^2} + m_0^2c^2\right)\left(\frac{\partial\chi}{\partial\varphi}\right)^2 -$$

$$-2\frac{e}{c}\left(\sum A_j\frac{\partial\chi}{\partial x_j} - \frac{1}{c^2}W\frac{\partial\chi}{\partial t}\right)\frac{\partial\chi}{\partial\varphi} = 0 .$$

We consider that φ is stationary, hence

$$\frac{\partial\chi}{\partial t} = E\frac{\partial\chi}{\partial\varphi} ,$$

and from this we obtain from the previous equation

$$\sum_j\left(\frac{\partial\chi}{\partial x_j}\right)^2 + \left[\frac{e^2}{c^2}\sum_j A_j^2 - m^2v^2\right]\left(\frac{\partial\chi}{\partial\varphi}\right)^2 - 2\frac{e}{c}\sum_j A_j\left(\frac{\partial\chi}{\partial x_j}\right)\frac{\partial\chi}{\partial\varphi} = 0,$$

which is the equation of the characteristics of the wave equation

$$\Delta\phi + \left[\frac{e^2}{c^2}\left(\sum_j A_j^2\right) - m^2v^2\right]\frac{\partial^2\phi}{\partial\varphi^2} - 2\frac{e}{c}\sum_j A_j \frac{\partial^2\phi}{\partial x_j \partial\varphi} = 0 .$$

The coefficients of a solution of the form

$$\phi = C\cos 2\pi\lambda\varphi + D\sin 2\pi\lambda\varphi$$

verify the equation

$$\Delta S = 4\pi\lambda^2\left[\frac{e^2}{c^2}\sum A_j^2 - m^2v^2\right]S - 4\pi\lambda\frac{e}{c}\sum A_j\frac{\partial S}{\partial x_j} = 0 .$$

The effective quantification of such an equation of the Schrödinger type is outside the purpose of this work.

We observe only, that the methods used in the invariantive mechanics lead to the Schrödinger equation.

5. The General Case. In Newtonian mechanics the general case is that of the fields of forces which do not necessarily derive from a potential. In this case it appears as a primary size which takes the place of the potential, or more exactly of its variation, the mechanical work represented by

$$X_1 dx_1 + X_2 dx_2 + X_3 dx_3$$

which is no longer necessarily an exact differential, as it usually is, if we have a potential.

In the invariantive mechanics as in the classical one, the potentials A_1, A_2, A_3, A_0 can not come directly, but by their variations

$$dA_j = A_{j1}dx^1 + A_{j2}dx^2 + A_{j3}dx^3 + A_{jo}dt \; ; \; j = 1,2,3,0 .$$

Whether there are potentials or not, the movement law comes out from equalizing the expression $D\omega_\delta^{(i)}$ with $D\omega_\delta^{(p)}$ given by the equality

$$D\omega_\delta^{(p)} = \sum_{j=1,2,3}\ \sum_{k=1,2,3,0} A_{jk}\, dx^k\, \delta x^j - \sum_{j=1,2,3}\ \sum_{k=1,2,3,0} A_{jk}\, dx^j\, \delta x^k +$$

$$+ \sum_{k=1,2,3,0} A_{ok}\, dt\, \delta x^k - \sum_{k=1,2,3,0} A_{ok}\, dx^k\, \delta t = \sum_{j=1,2,3}\left[\sum_{k=1,2,3,0} (A_{jk} - A_{kj}) + (A_{jo} + A_{oj})dt\right] \delta x^j -$$

$$\sum_{j=1,2,3} (A_{jo} - A_{oj})\, dx^j \quad \delta t \ .$$

Taking into account that

$$D\omega_\delta^{(i)} = \sum \dot{p}_j\, \delta x^j - H\delta t,$$

where $p_j = mx_j$; $H = mc^2$, we obtain the equations of the movement under the form

$$\frac{dp_j}{dt} = \sum_{k=1,2,3} (A_{jk} - A_{kj})\dot{x}^k + A_{jo} - A_{oj}$$

$$\frac{dH}{dt} = \frac{1}{c^2} \sum_{j=1,2,3} (A_{jo} - A_{oj})\dot{x}^j \ .$$

The place of the force from classical Mechanics is taken in Invariantive Mechanics by the tensor A_{jk} ($j,k \check{=} 1,2,3,0$) with 16 components. The only limitation could result from the compatibility conditions of these four equations, taking into account that $H = mc^2$. But one finds out immediately, because these conditions are identically verified, given the Lorentzian expression of the mass.

It comes out that we have no a priori limitation for the values of the field tensor with 16 components A_{jk}.

If the spatial part of the tensor is symmetric, that is if

$$A_{jk} = A_{kj} \ ; \ j,k = 1,2,3$$

and if the temporal components A_{jo} are null, then the motion equations become like the Newtonian ones

$$\dot{p}_j = A_{oj} \ ; \ j = 1,2,3.$$

§ 5. Anti-Minkowskian Universes

1. On the alternative $\epsilon = \pm 1$. The interpretation as a fibre of the Einstein-Minkowski space makes more accessible the notion of other forms of fibres which are required as a mathematical necessity by the theory of mechanics expounded in the first paragraph of this chapter.

In the development of the axiomatic process of the mechanics of inertial motion, after stating in the first postulate that the ratio E/m represents a universal constant which leads to $E = \epsilon m\omega^2$, we have further to decide whether ϵ is equal to $+1$ or to -1. In the second postulate, which we have called the Einstein-Minkowski postulate, ϵ is equal to $+1$ in accordance with the current experience where velocities higher than the velocity of light are not encountered, hence where the velocities are bounded.

The second possibility, represented by $\epsilon = -1$, cannot be excluded on empirical grounds since we cannot exclude possible experiments, forms of matter and physical fields where the velocities may reach arbitrary values and where ω, which represents the universal constant of the first postulate, may have a different meaning than that of a limiting velocity. It would not be the first time when a mathematical theory leads to such a situation. A similar case occurred with Dirac's theory which opened the way to the investigations concerning the existence of the positron.

Therefore we must consider also the case $\epsilon = -1$ with all the theoretical consequences of the respective Mechanics.

2. The Inertial Mechanics of the Postulate $\epsilon = -1$. In that case we have

$$m = m_0 \Big/ \sqrt{1 + \frac{v^2}{\omega^2}} \quad ; \quad E = - m\omega^2 \tag{50}$$

and we do not know of any fact, either experimental or theoretical, indicating a preferential value for ω. But the expression of m, which may be written

$$m = \sqrt{m_0^2 - \frac{1}{\omega^2}\left(p_1^2 + p_2^2 + p_3^2\right)},$$

decreases when the velocity increases, the mass at rest being its greatest mass,

$$m \leqslant m_0.$$

This inertial mass, whose existence is so far purely mathematical, and which behaves in accordance with the first of the relations (50), will be called anti-mass of the ω species.

The four-dimensional fibre-universe associated with such a mass derives evidently from the form

$$\omega_\delta^{(i)} = \sum_{j=1,2,3} p_j\, \delta x_j + m\omega^2\, \delta t$$

hence it corresponds to the Euclidean quadratic form

$$d\theta^2 = dx_1^2 + dx_2^2 + dx_3^2 + \omega^2 dt^2$$

and may be called anti-Minkowkian, anti-Einsteinean.

It is clear, that the anti-Einsteinean fibres, possessing the structure of a Euclidean space, are simpler from a geometrical point of view. In such fibres there are no paths of zero length as in the case of Einstein fibres when it is possible to have

$$d\sigma^2 = c^2 dt^2 - dx_1^2 - dx_2^2 - dx_3^2 = 0$$

for the motion of a photon whose velocity is c, and moreover, where there are no regions forbidden to motion as in the Einstein fibre.

The anti-Einsteinean fibres are homogeneous Euclidean universes different from one another according to the values of ω; this introduces also a certain dissimilarity between these spaces, provided, however, that the time unit is physically defined.

3. Anti-Maxvellian Field. The structure of $\omega_\delta^{(i)}$ has imposed a definite form for structural equations of the field potentials $\tilde{A}(A, A_0)$.

In the stationary case these potentials must satisfy an equation of the hyperbolic type

$$\Delta U - \frac{1}{c^2}\,\frac{\partial^2 U}{\partial t^2} = 0\,, \tag{51}$$

which is essentially an equation of propagation specific of the Maxwellian field.

In the case of a variable field, the potentials $A(A_1, A_2, A_3)$, A_0, must satisfy equations of the form

$$\Delta \mathbf{U} - \frac{1}{c^2}\,\frac{\partial^2 \mathbf{U}}{\partial t^2} = 4\pi \mathbf{k}\,,$$

$$\Delta U_0 - \frac{1}{c^2}\,\frac{\partial^2 U}{\partial t^2} = 4\pi\theta\,,$$

where $\mathbf{U}$ corresponds to the vector A_1, A_2, A_3; U_0 corresponds to A_0, $\mathbf{k}$ is a time vector and θ is a scalar.

If the inertial motion derives from the Anti-Einsteinean form and belongs to a fibre of metrics $\epsilon = -1$, then the four potentials of the field verify, in the stationary case, the equations

$$\operatorname{div} \mathbf{A} + \frac{1}{\omega}\,\frac{\partial A_0}{\partial t} = 0\,,$$

$$\Delta A_j + \frac{1}{\omega^2}\,\frac{\partial^2 A_j}{\partial t^2} = 0\;;\quad j = 1,2,3,0$$

and the two fields $\mathbf{E}_\omega$ and $\mathbf{H}_\omega$ defined in E_3 by the relations

$$\mathbf{E}_\omega = \operatorname{grad} A_0 - \frac{\partial \mathbf{A}}{\partial t}\,,$$

$$\mathbf{H}_\omega = \operatorname{rot} \mathbf{A}$$

(where $\mathbf{A}$ is the vector A_1, A_2, A_3) verify the two groups of anti-Maxwellian equations

$$\operatorname{rot} \mathbf{E}_\omega + \frac{\partial \mathbf{H}_\omega}{\partial t} = 0\,;\quad \operatorname{div} \mathbf{H}_\omega = 0$$

and

$$\text{rot } \mathbf{H}_\omega - \frac{1}{\omega^2} \frac{\partial \mathbf{E}_\omega}{\partial t} = 0 \,, \tag{52}$$

$$\text{div } \mathbf{E}_\omega = 0 \,,$$

where the first group is identical to the Maxwellian group, while the second is essentially different.

The essential difference between the two fields is manifest. Whereas equation (51) corresponds to a propagation process in the Einstein fibre and equations (52) correspond to another type of physical phenomenon, we shall call it a phenomenon of global inner stress.

In this case too, one may construct a variable field by introducing a vector $\tilde{k}(\mathbf{k}, k_0)$ in the four-dimensional space and such that the potentials A verify the equation

$$\Delta\tilde{A} + \frac{1}{\omega^2} \frac{\partial^2 \tilde{A}}{\partial t^2} = - 4\pi\tilde{k} \ .$$

<u>Conclusions and Problems.</u> There are various interpetations in phenomenological terms of the previous model of the universe. There is nothing against admitting a matter whose behaviour should be in accordance with formulas (50) and a field or fields of the type (50), (51), (52) with a possible charge corresponding to the electric charge in a Maxwell field.

It remains to be seen, whether an antiparticle behaves in the way indicated by these formulas.

CHAPTER II

THE MECHANICS OF STABLE PARTICLES

Introduction

The motion of a rigid body, writes T. Levi-Civita in his Kinematics, extends naturally to the whole space and may be studied without specifying the particular reference system used for defining it. It is a motion with six degrees of freedom. The velocity of a generic point P of a rigid space in motion is given by the expression $\mathbf{v}_p = \mathbf{v}_o + \omega \times \mathbf{r}$ where $\mathbf{v}_o$ is the velocity of a point O of the system, ω is the angular velocity common to all the points of the space and $\mathbf{r}$ is the vector **OP**. It is obvious that $\mathbf{v}_p$ increases indefinitely with the distance r, in contradiction with present day physics, by which there exists a limiting velocity.

Hence, the concept of a rigid system unlimited in space cannot be transposed in invariant mechanics. The extension of systems in space must necessarily intervene in the definition of any material system in motion, at least as the system will be characterized with the help of the moments of inertia. In the calculation of the respective moments we have to consider the distribution of mass which, being dependent on velocity, should not present important variations caused by differences in velocity, hence by the extension of the body.

It follows that the bodies which we may consider as rigid are of limited extension and, since in the case of current dimensions and velocities, the invariant mechanics should reduce to Newton's mechanics, we shall consider first particles whose small dimensions make possible a theory which proves to be, however, quite different from that of Newton.

§1. Inertial Motion

1. Euclidean Invariants. The position of a body in the three-dimensional Euclidean representative space is defined by a point $P(x_1, x_2, x_3)$, which may be the center of mass of the body, and by three angles $\alpha_1, \alpha_2, \alpha_3$. The dynamic magnitudes which must be associated with the position P and the angles $\alpha_1, \alpha_2, \alpha_3$ are evidenced by performing the calculation for the construction of the inertial form $\Omega_\delta^{(i)}$ of the respective body.

The spatial part of the form $\Omega_\delta^{(i)}$, which will be denoted by Π_δ, is represented by the integral

$$\Pi_\delta = \int_c (\dot{\xi}_1 \delta\xi_1 + \dot{\xi}_2 \delta\xi_2 + \dot{\xi}_3 \delta\xi_3) dm, \tag{1}$$

which is the integral of the spatial part of the form $\Omega_\delta^{(i)}$:

$$\Omega_\delta^{(i)} = \int_c \omega_\delta^{(i)} = \int_c \sum_j^{1,2,3} (\dot{\xi}_j \delta\xi_j - H\delta t) dm. \tag{2}$$

The vector $\delta\xi(\delta\xi_1, \delta\xi_2, \delta\xi_3)$ corresponds to a rigid displacement of the system, hence

$$\delta\xi = \delta\mathbf{x} + \mathbf{r} \times \delta\alpha, \tag{3}$$

where $\mathbf{r} = \xi - \mathbf{x}$, the components of which are $\zeta_1, \zeta_2, \zeta_3$.

This makes evident the angles $\delta\alpha(\delta\alpha_1, \delta\alpha_2, \delta\alpha_3)$, which together with $\mathbf{x}(x_1, x_2, x_3)$ characterize the motion.

Using these expressions also for $\delta = d$ in (1), we obtain

$$\begin{aligned} \Pi_\delta = m \sum x_j \delta x_j &+ \left[(\ell_{22} + \ell_{33})\dot{\alpha}_1 - \ell_{12}\dot{\alpha}_2 + \ell_{13}\dot{\alpha}_3\right] \delta\alpha_1 + \\ &+ \left[(\ell_{33} + \ell_{11})\dot{\alpha}_2 - \ell_{23}\dot{\alpha}_3 + \ell_{21}\dot{\alpha}_1\right] \delta\alpha_2 + \left[(\ell_{11} + \ell_{22})\dot{\alpha}_3 - \ell_{31}\dot{\alpha}_1 + \ell_{32}\dot{\alpha}_2\right] \delta\alpha_3, \end{aligned} \tag{4}$$

where

$$m = \int_c dm \tag{5}$$

is the total mass of the body and

$$\ell_{jk} = \int_c \zeta_j \zeta_k \, dm$$

are the moments of the second order of the mass distribution of the body, considering that $\zeta_1, \zeta_2, \zeta_3$, are the components of $\mathbf{r} = \mathbf{OP}$ in a reference frame rigidly attached to c; hence ℓ_{jk} are values depending on the body and not on position. The expressions

$$\begin{aligned} \theta_1 &= (\ell_{22} + \ell_{33})\dot{\alpha}_1 - \ell_{12}\dot{\alpha}_2 + \ell_{13}\dot{\alpha}_3 \\ \theta_2 &= \ell_{21}\dot{\alpha}_1 + (\ell_{33} + \ell_{11})\dot{\alpha}_2 - \ell_{23}\dot{\alpha}_3 \\ \theta_3 &= -\ell_{31}\dot{\alpha}_1 + \ell_{32}\dot{\alpha}_2 + (\ell_{11} + \ell_{22})\dot{\alpha}_3 \end{aligned} \tag{6}$$

are the angular impulses of motion; therefore the inertial form of the body may be written

$$\Omega_\delta^{(i)} = \sum_{j=1,2,3} m\dot{x}_j\,\delta x_j + \sum_{j=1,2,3} \theta_j\,\delta\alpha_j - H\delta t, \tag{7}$$

where we have to specify the dependence of m and θ on the other elements of the motion as well as the expression of H. To this end it is necessary to state precisely which are the Euclidean invariants of the motion, except $\alpha = 1/2\,(p_1^2 + p_2^2 + p_3^2)$, where $p_j = m\dot{x}_j$, and $1/2\ \theta_j^2 (j = 1,2,3)$ which are evidently invariant.

We obtain these invariants first by reversing the relations (6). We have

$$\begin{aligned} \dot{\alpha}_1 &= \varkappa_1\theta_1 - \lambda_3\theta_2 + \lambda_2\theta_3, \\ \dot{\alpha}_2 &= \lambda_3\theta_1 + \varkappa_2\theta_2 - \lambda_1\theta_3, \\ \dot{\alpha}_3 &= -\lambda_2\theta_1 + \lambda_1\theta_2 + \varkappa_3\theta_3, \end{aligned} \tag{8}$$

where $\varkappa_j (j = 1,2,3)$ and $\lambda_j (j = 1,2,3)$ are rational expressions of ℓ_{jk}.

Obviously, only compatible systems are considered, that is, systems for which the inversion (6) to (8) is possible.

From (8) it follows immediately the expression

$$\begin{aligned} \dot{\alpha}_1\,\delta\theta_1 + \dot{\alpha}_2\,\delta\theta_2 + \dot{\alpha}_3\,\delta\theta_3 &= \sum \varkappa_j\,\delta(1/2\,\theta_j^2) + \\ &+ \lambda_1(\theta_2\delta\theta_3 - \theta_3\delta\theta_2) + \lambda_2(\theta_3\delta\theta_1 - \theta_1\delta\theta_3) + \lambda_3(\theta_1\delta\theta_2 - \theta_2\delta\theta_1) \end{aligned} \tag{9}$$

which has an invariant geometrical character and evidences besides the invariants

(10) $$\omega_j = 1/2\ \theta_j^2\ ; \quad j = 1,2,3$$

also three non-holonomic invariants defined by the expressions

(11) $$\delta\varphi_1 = \theta_2\,\delta\theta_3 - \theta_3\,\delta\theta_2\,, \quad \delta\varphi_2 = \theta_3\,\delta\theta_1 - \theta_1\,\delta\theta_3\,, \quad \delta\varphi_3 = \theta_1\,\delta\theta_2 - \theta_2\,\delta\theta_1\ .$$

The function of state H is, in principle, a function of the Euclidean invariants of the system:

$$H = H(\alpha\,,\ \omega_1\,,\omega_2\,,\omega_3\,,\ \varphi_1\,,\varphi_2\,,\varphi_3)$$

occurring as dH or δH which may be effectively written.

2. The Law of Motion. The law of motion is expressed by the condition

$$D\Omega_\delta^{(i)} = d\Omega_\delta^{(i)} - \delta\Omega_d^{(i)} = 0$$

for any δ . Obviously we have

(12) $$D\Omega_\delta^{(i)} = \sum dp_j\,\delta x_j + \sum d\theta_j\,\delta\alpha_j - dH\delta t - \sum dx_j\,\delta p_j - \sum d\alpha_j\,\delta\theta_j + \delta H dt\ .$$

By making the coefficients of δx_j, $\delta\alpha_j$, δt equal to zero and since H does not include, by hypothesis, the variables x, α and t, but only the invariants quoted, we obtain the first group of equations

(13) $$\frac{dp_j}{dt} = 0,\ \frac{d\theta_j}{dt} = 0,\ \frac{dH}{dt} = 0 \qquad (j = 1, 2, 3)\ ;$$

which prove the uniformity of the motion and the conservation of energy H. The second group of equations is obtained by cancelling the coefficients of δp_j and $\delta\theta_j$. We have, then, in the first place

$$\frac{dx_j}{dt} = \frac{\partial H}{\partial p_j} \quad ; \; j = 1, 2, 3 \; .$$

Since H is dependent on p_1, p_2, p_3 through the agency of α, the above equations give

$$\dot{x}_j = \frac{\partial H}{\partial \alpha} \, p_j \qquad j = 1, 2, 3 \tag{14}$$

and putting

$$m = 1/\frac{\partial H}{\partial \alpha}$$

we obtain the impulse

$$p = mv \; (p_j = m\dot{x}_j^1 \; , \; j = 1, 2, 3) \; . \tag{15}$$

The second group of equations, resulting from the cancellation of the coefficients of the variations $\delta\theta_1, \delta\theta_2, \delta\theta_3$, is of the form

$$\dot{\alpha}_j = \frac{\partial H}{\partial \theta_j} \, , \tag{16}$$

whence, using (9), there follows

$$\sum_j \frac{\partial H}{\partial \theta_j} \, \delta\theta_j = \sum_j \varkappa_j \, \delta\omega_j + \sum_j \lambda \, \delta\varphi_j \; . \tag{17}$$

Since, through the agency of the invariants ω and φ, H is dependent on $\theta_1, \theta_2, \theta_3$, and the first member of (17) may be written

$$\sum_j \frac{\partial H}{\partial \omega_j} \, \delta\omega_j + \sum_j \frac{\partial H}{\partial \varphi_j} \, \delta\varphi_j \; ,$$

hence we may write

$$\varkappa_j = \frac{\partial H}{\partial \omega_j} \; ; \; \lambda_j = \frac{\partial H}{\partial \varphi_j} \; ; \quad (j = 1, 2, 3.) \; .$$

3. Two Conventions. In order to continue, two further conventions which characterize the (rigid) body, are required.

1st . Convention. Obviously the tensor

$$\epsilon_{jk} = \frac{1}{m} \, \ell_{jk} \; ; \quad j,k = 1, 2, 3$$

has the dimension of the squared length. We assume that it does not depend on the mass distribution as in the case of the classical homogeneous rigid body. Then equations (6) may be written also

$$\theta_1 = (\epsilon_{22} + \epsilon_{33}) m\dot{\alpha}_1 - \epsilon_{12} m\dot{\alpha}_2 + \epsilon_{13} m\dot{\alpha}_3 \, , \text{ etc.} \tag{18}$$

Applying to (18) the inversion which has led from (6) to (8), we obtain the relations

$$\begin{aligned} m\dot{\alpha}_1 &= u_1 \theta_1 - v_3 \theta_2 + v_2 \theta_3 \, , \\ m\dot{\alpha}_2 &= v_3 \theta_1 + u_2 \theta_2 - v_1 \theta_3 \, , \\ m\dot{\alpha}_3 &= - v_2 \theta_1 + v_1 \theta_2 + u_3 \theta_3 \, , \end{aligned} \tag{19}$$

where $u_1, u_2, u_3 ; v_1, v_2, v_3$ depend only on the spatial structure of the body.

2nd .Convention. The assumption that the relation between the mass of the material body in inertial motion, which is a constant, and the energy of the same body throughout the motion, which is also constant, be independent of the initial conditions of the motion, is necessary. If it were not independent, contradictions would occur which temporarily shall not be looked into.

If we have $\dot{\alpha}_j^0 = 0 \, (j = 1, 2, 3)$, relations (8) show that $\theta_j^0 = 0 \, (j = 1, 2, 3)$, and equations (13) show that during the motion $\theta_j = 0 \, (j = 1, 2, 3)$. The body behaves like a material particle and we have

$$H = c^2 m \tag{20}$$

hence, the relation holds for a rigid body for any initial conditions.

We can now resume the examination of the motion starting from the form

$$\Omega_\delta^{(i)} = \sum_j p_j \delta x_j + \sum_j \theta_j \delta\alpha_j - mc^2 \delta t \ . \tag{21}$$

Taking into account the relations obtained above, it follows that

$$D\Omega_\delta^{(i)} = -\sum_j dx_j \delta p_j - \sum_j d\alpha_j \delta\theta_j + c^2 dt\, \delta m = 0 \ .$$

Multiplying by m, we obtain the relation

$$\sum_j p_j \delta p_j + \sum_j m\dot{\alpha}_j \delta\theta_j = c^2 m \delta m \ . \tag{22}$$

If we use (19), we have the equality

$$\delta(1/2 \sum_j p_j^2 + 1/2 \sum u_j \theta_j^2 + \sum v_j \varphi_j) = 1/2\ c^2 \delta(m^2),$$

since u_j and v_j are independent of θ_j and α_j.

Hence, with the notations used before, there follows that

$$\alpha + \sum_j u_j \omega_j + \sum v_j \varphi_j = 1/2\ c^2 (m^2 - m_o^2), \tag{23}$$

where m_o is the mass of the body at rest ($\mathbf{p} = 0$, $\theta = 0$).

From (23) we obtain the expression of the dynamical mass

$$m = \sqrt{m_o^2 + 2/c^2 (\alpha + \sum_j u_j \omega_j + \sum_j v_j \varphi_j)} \tag{24}$$

and of the energy

$$H = c \sqrt{m_o^2 c^2 + 2(\alpha + \sum_j u_j \omega_j + \sum_j v_j \varphi_j)} \ .$$

Since $\delta\varphi$ has the form of a vectorial product of $\theta(\theta_1, \theta_2, \theta_3)$, and $\delta\theta(\delta\theta_1, \delta\theta_2, \delta\theta_3)$, $\delta\varphi = \theta \times \delta\theta$, we obtain for a motion where θ varies as a function of t the expression

$$\varphi = \varphi_0 + \int_0^t \theta(s) \times \dot{\theta}(s) \, ds,$$

where $\theta(t)$ corresponds to the effective law of motion.

§ 2. Motion of the Stable Particles in a Field

1. The Potential Form. $\Omega_\delta^{(p)}$. In accordance with the inertial form (21, §1) of $\Omega_\delta^{(i)}$ the field is defined by two vector potentials $\mathbf{A}(A_1, A_2, A_3)$ and $\mathbf{B}(B_1, B_2, B_3)$ relative to the position of P and to the body orientation respectively, and by a scalar potential C.

Hence we have

$$\Omega_\delta^{(p)} = \sum A_j \delta x_j + \sum B_j \delta\alpha_j - C\delta t \ . \tag{1}$$

A priori, we must consider **A, B** and C as functions of all the variables $x_1, x_2, x_3, \alpha_1, \alpha_2, \alpha_3$, t. However, if we confine ourselves to the classical fields and adopt for **A, B** and C the calculations indicated by the nature of these fields – as we shall do in the next paragraph – then **A, B** and C must be considered as functions of x_1, x_2, x_3 and t only.

Note: Nevertheless, we cannot exclude the existence of fields, irreducible to elementary components, as will be assumed in the next paragraph, and hence the possible dependency of the fields **A, B,** C on the angles α as well.

For these reasons we shall first perform the calculation in the general case.

2. Equation of Motion. We have evidently

$$D\Omega_\delta^{(p)} = \sum dA_j \delta x_j + \sum dB_j \delta\alpha_j - dC\delta t - \sum dx_h \delta A_h - \sum d\alpha_h \delta B_h + dt\delta C \tag{2}$$

or, writing explicitly,

$$\begin{aligned} D\Omega_\delta^{(p)} = & \sum \Big[\Big(dA_j - \sum dx_h \frac{\partial A_h}{\partial x_j} - \sum d\alpha_h \frac{\partial B_h}{\partial x_j} + dt \frac{\partial C}{\partial x_j}\Big) \delta x_j + \\ & + \Big(dB_j - \sum dx_h \frac{\partial A_h}{\partial \alpha_j} - \sum d\alpha_h \frac{\partial B_h}{\partial \alpha_j} + dt \frac{\partial C}{\partial \alpha_j}\Big) \delta\alpha_j \Big] - \\ & - \Big(dC + \sum dx_h \frac{\partial A_h}{\partial t} + \sum d\alpha_h \frac{\partial B_h}{\partial t} - dt \frac{\partial C}{\partial t}\Big) \delta t \ . \end{aligned} \tag{3}$$

On the other hand by expanding (21) we obtain

$$D\Omega_\delta^{(i)} = \sum dp_j \delta x_j + \sum d\theta_j \delta\alpha_j - dH\delta t -$$

$$-\left(dx_j - \frac{\partial H}{\partial p_j}\right)\delta p_j - \left(d\alpha_j - dt\,\frac{\partial H}{\partial \theta_j}\right)\delta\theta_j \;. \tag{4}$$

Or the relations

$$\frac{dx_j}{dt} = \frac{\partial H}{\partial p_j} \;;\; \frac{d\alpha_j}{dt} = \frac{\partial H}{\partial \theta_j} \;;\; j = 1, 2, 3 \tag{5}$$

are identically satisfied by the definition of impulses, which have been already obtained in §1.

We further obtain the equations of motion by writing $D\Omega_\delta^{(i)} = D\Omega_\delta^{(p)}$, which gives

$$\frac{dp_j}{dt} = \frac{dA_j}{dt} - \sum \dot{x}_h \frac{\partial A_h}{\partial x_j} - \sum \dot{\alpha}_h \frac{\partial B_k}{\partial x_j} + \frac{\partial C}{\partial x_j}\,, \tag{6}$$

$$\frac{d\theta_j}{dt} = \frac{dB_j}{dt} - \sum \dot{x}_h \frac{\partial A_h}{\partial \alpha_j} - \sum \dot{\alpha}_h \frac{\partial B_h}{\partial \alpha_j} + \frac{\partial C}{\partial \alpha_j} \tag{7}$$

$$j = 1, 2, 3$$

and the equation of the energy exchange between body and field

$$\frac{dH}{dt} = \frac{\partial C}{\partial t} - \frac{dC}{dt} - \sum \dot{x}_h \frac{\partial A_h}{\partial t} - \sum \dot{\alpha}_h \frac{\partial B_h}{\partial t}\,. \tag{8}$$

The first group of equations may be written also

$$\begin{aligned}\frac{dp_1}{dt} = {} & \frac{\partial A_1}{\partial t} - \frac{\partial C}{\partial x_1} + \left(\frac{\partial A_1}{\partial x_2} - \frac{\partial A_2}{\partial x_1}\right)\dot{x}_3 - \left(\frac{\partial A_3}{\partial x_1} - \frac{\partial A_1}{\partial x_3}\right)\dot{x}_2 + \\ & \left(\frac{\partial A_1}{\partial \alpha_1} - \frac{\partial B_1}{\partial x_1}\right)\dot{\alpha}_1 + \left(\frac{\partial A_1}{\partial \alpha_2} - \frac{\partial B_2}{\partial x_2}\right)\dot{\alpha}_2 + \left(\frac{\partial A_1}{\partial \alpha_3} - \frac{\partial B_3}{\partial x_3}\right)\dot{\alpha}_3\end{aligned} \tag{9}$$

together with two other similar equations.

Hence, there exist an electric field, a magnetic field and a mixed field which together determine the motion of the body.

Special case where **A,B,C** are dependent only on x_1, x_2, x_3 and t. In that case the equations of motion are considerably simplified. In vectorial notation, equation (6) becomes

$$\frac{d\mathbf{p}}{dt} = \mathbf{E} + \mathbf{v} \times \mathbf{H} - \text{grad } K, \tag{10}$$

where

$$\begin{aligned} \mathbf{E} &= \frac{\partial \mathbf{A}}{\partial t} + \text{grad } C, \\ \mathbf{H} &= \text{rot } \mathbf{A}, \\ K &= \alpha_1 B_1 + \alpha_2 B_2 + \alpha_3 B_3 \ . \end{aligned} \tag{11}$$

Equation (7) reduces to

$$\frac{d\theta_j}{dt} = \frac{dB_j}{dt} \ ; \quad j = 1, 2, 3 \tag{12}$$

hence in vectorial notation

$$\boldsymbol{\theta} - \boldsymbol{\theta}_o = \mathbf{B} - \mathbf{B}_o \ .$$

3. Computation of the Elements **A, B, C**. The current calculation method of the form $\Omega_\delta^{(p)}$ with respect to the body c results from the hypothesis of the additivity of the elementary forms

$$\omega_\delta^{(p)} = \mathbf{a}(\xi, t)\delta\xi - a_o(\xi, t)\delta t \tag{13}$$

corresponding to the elementary components of the body whose mass is $d\mu$ and which may be of an arbitrary nature, e.g. an electric charge or any other unspecified type of charge or mass.

Therefore, by definition, we have

$$\Omega_\delta^{(p)} = \int_c (\mathbf{a}(\xi,t)\delta\xi - a_o(\xi,t)\delta t) d\mu \tag{14}$$

and we put

$$\mu = \int_c d\mu$$

with the implicit assumption that μ be finite.

We also put

$$\xi = \mathbf{x}_g + \mathbf{r} \tag{15}$$

where $\mathbf{x}$ is the position vector of the center of mass G of the body and $\mathbf{r}$ is the vector GP. For any displacement of the rigid body we have

$$d\xi = \delta \mathbf{x}_g + \mathbf{r} \times \delta\alpha \ . \tag{16}$$

Hence

$$\begin{aligned} \Omega_\delta^{(p)} = & \left[\int_c \mathbf{a}(\mathbf{x}_g + \mathbf{r}, t)\, \delta\mu\right] \delta\mathbf{x}_g + \left[\int_c (\mathbf{a}(\mathbf{x}_g + \mathbf{r}, t) \times \mathbf{r})\, d\mu\right] \delta\alpha - \\ & - \left[\int_c a_o(\mathbf{x}_g + \mathbf{r}, t)\, d\mu\right] \delta t \ . \end{aligned} \tag{17}$$

The coefficients of $\delta\mathbf{x}_g$, $\delta\alpha$, and δt are evidently dependent on $\mathbf{x}_g$ and t.

The integrals refer to r and t, and do not depend on the coordinates. They shall be computed with respect to a reference system rigidly attached to the body and be independent of all elements associated with the motion, other than $\mathbf{x}_g$ and t. Hence, we shall have in the general case starting from the formula (14):

$$\Omega_\delta^{(p)} = \mathbf{A}(\mathbf{x}_g, t)\, \delta\mathbf{x}_g + \mathbf{B}(\mathbf{x}_g, t)\, \delta\alpha - C(\mathbf{x}_g, t)\, \delta t \ . \tag{18}$$

4. The Field of Particles. In the case in which we are particularly interested, namely that of particles of dimensions such that the previous approximations are permitted, we may assume that the field functions $\tilde{a}(\xi, t)$ have continuous partial differential coefficients up to the second order. For the representation of the position r we consider a reference system attached to the body; then the components of r will be ζ_h (h = 1,2,3) and we may write

$$\begin{aligned} \tilde{a}(\mathbf{x} + \mathbf{r}, t) = & \ \tilde{a}(\mathbf{x}, t) + \sum_h \zeta_h \tilde{a}_h(\mathbf{x}, t) + 1/2 \sum_{h,k} \zeta_h \zeta_k \tilde{a}_{hk}(\mathbf{x}, t) + \\ & + 1/2 \sum_{h,k} \zeta_h \zeta_k \left[\tilde{a}_{hk}(x_1 + u_1\zeta_1, x_2 + u_2\zeta_2, x_2 + u_3\zeta_3, t) - \tilde{a}_{hk}(\mathbf{x}, t)\right], \end{aligned} \tag{19}$$

where $|u_1|, |u_2|, |u_3| < 1$.

There follows

$$\tilde{A}(x,t) = \mu\tilde{a}(x,t) + \sum_j \ell_j \tilde{a}_j(\mathbf{x},t) + 1/2 \sum_{j,k} \ell_{jk}\, \tilde{a}_{jk}(x,t) + \epsilon, \tag{20}$$

where

$$\ell_j = \int_c \zeta_j d\mu, \quad \ell_{jk} = \int_c \zeta_j \zeta_k \, d\mu$$

$$\epsilon = \sum_{j,k} \int_c \zeta_j \zeta_k \left[\tilde{a}_{jk}(x_1 + u_1\zeta_1, x_2 + u_2\zeta_2, x_3 + u_3\zeta_3, t) - \tilde{a}_{jk}(x_1, x_2, x_3, t)\right] d\mu$$

and

$$\tilde{a}_j = \frac{\partial \tilde{a}}{\partial x_j}, \quad \tilde{a}_{jk} = \frac{\partial^2 \tilde{a}}{\partial x_j \partial x} \quad ; \quad j,k = 1, 2, 3$$

$\tilde{a}$ being the vector (a_1, a_2, a_3, a_0).

By the same conventions and taking into account (18) we have

$$\begin{aligned} \mathbf{B}(\mathbf{x}, t) &= \int_c \left[\, \mathbf{a}(\mathbf{x}+\mathbf{r},t) \times \mathbf{r}\right] d\mu = \\ &= \mathbf{a}(\mathbf{x},t) \times \mathbf{I} + \sum_j (\mathbf{a}_j(\mathbf{x},t) \times \mathbf{I}_j) + \varkappa, \end{aligned} \tag{21}$$

where $\mathbf{I}$ is the vector of components $\int_c \zeta_h d\mu$ $(h = 1, 2, 3)$ and $\mathbf{I}_j$ the vector system of components $\int_c \zeta_j \zeta_h d\mu$ $(h = 1, 2, 3)$, $\varkappa$ consisting of a number of terms of the form

$$\int_c \zeta_j \zeta_h \zeta_k \, b(\zeta,t)\, d\mu$$

of an order lower than $K\lambda^3$ and which are assumed to be negligible. Therefore we have

$$\begin{aligned} \Omega_\delta^{(p)} &= (\mu\mathbf{a} + \sum_j \ell_j \mathbf{a}_j + 1/2 \sum_{j,k} \ell_{jk} \mathbf{a}_{jk} + \epsilon)\, \delta\mathbf{x} + \\ &+ \left[\mathbf{a} \times \mathbf{I} + \sum_j (\mathbf{a}_j \times \mathbf{I}_j) + \varkappa\right] \delta\alpha - \left[\mu a_o + \sum_j \ell_j a_{oj} + \sum_{j,k} I_{jk}\, a_{ojk} + \epsilon_o\right] \delta t, \end{aligned} \tag{22}$$

where ℓ_j and ℓ_{jk} are invariant characteristics of the body C and $\mathbf{a}$ together with its derivatives a_j, a_{jk} are continuous functions depending only on $\mathbf{x}$ and t.

5. Classification of Particles. A particle is said to be of the first species if the terms containing ℓ_{jk} are negligible. In that case we have

$$\Omega_\delta^{(p)} = (\mu \mathbf{a} + \sum \ell_j \mathbf{a}_j)\,\delta \mathbf{x} + (\mathbf{a} \times \mathbf{I})\,\delta\alpha - (\mu a_o + \sum \ell_j a_{oj})\,\delta t\,. \tag{23}$$

The particle is said to be of the second species if the terms whose order is higher than the stresses ℓ_{jk} are negligible, and we have

$$\begin{aligned}\Omega_\delta^{(p)} = {} & (\mu \mathbf{a} + \sum \ell_j \mathbf{a}_j + 1/2 \sum \ell_{jk} \mathbf{a}_{jk})\,\delta \mathbf{x} + \\ & [\mathbf{a} \times \mathbf{I} + \sum (\mathbf{a}_j \times \mathbf{I}_j)]\,\delta\alpha - (\mu a_o + \sum \ell_j a_{oj} + 1/2 \sum \ell_{jk} a_{ojk})\,\delta t\,.\end{aligned} \tag{24}$$

If we assume that the magnitudes $\mathbf{a}$ and a_o are of a current order with respect to the unit charge, say of the same order in the case of an elementary particle, then, noting by e the unit charge, we have

$$\mu = \text{multiple of e}, \quad \ell_j \sim \mu.10^{-23}, \quad \ell_{jk} \sim \mu.10^{-46}\,.$$

If instead of a charge we have to deal with a mass, a mezon mass for instance, then we must assign to μ corresponding values which should not greatly affect the order of magnitude of ℓ_j or of ℓ_{jk}.

Hence the equation of motion of a particle of the first species are derived from (22), and we have also

$$\begin{aligned} H_I &= (\mathbf{a} \times \mathbf{I}).\alpha \\ \theta_I &= \theta_I^0 + [\mathbf{a}(\mathbf{x},t) - \mathbf{a}(\mathbf{x},0)] \times \ell\,. \end{aligned} \tag{25}$$

For a particle of the second species we have

$$H_{II} = [\mathbf{a} \times \mathbf{I} + \sum (\mathbf{a}_j \times \mathbf{I}_j)]\,\alpha\,,$$

and

$$\theta_{II} = \theta_{II}^0 + [\mathbf{a}(\mathbf{x},t) - \mathbf{a}(\mathbf{x},0)] \times \ell + \sum (\mathbf{a}_j(\mathbf{x},t) - \mathbf{a}_j(\mathbf{x},0)) \times \ell_j\,. \tag{26}$$

We must add to the equations (21) and (23) the equations

$$\frac{dx_j}{dt} = \frac{\partial H}{\partial p_j} \; ; \; \frac{d\alpha_j}{dt} = \frac{\partial H}{\partial \theta_j} \; ; \; (j = 1, 2, 3)$$

where H is given by (25) implying the presence of the values φ_j which do not vanish as in the particular case of inertial motion.

Commentary

The spatial part of the form $\Omega_\delta^{(i)}$ which we shall denote by Π_δ is represented by the integral

$$\Pi_\delta = \int_c \dot{\xi} . \delta\xi \, dm,$$

where $\dot{\xi} \, \delta\xi dm$ is the spatial part of the form $\omega_\delta^{(i)}$.

If $\mathbf{x}(x_1, x_2, x_3)$ corresponds to the center of gravity of the body at the time t and putting $\xi = \mathbf{x} + \mathbf{r}$, we have

$$\dot{\xi} = \dot{\mathbf{x}} + \mathbf{r} \times \dot{\alpha} ,$$

where $\dot{\alpha}$ is a vector, the components of which are $\dot{\alpha}_1, \dot{\alpha}_2, \dot{\alpha}_3$.

Similarly we have

$$\delta\xi = \delta\mathbf{x} + \mathbf{r} \times \delta\alpha ,$$

whence

$$\dot{\xi}\delta\xi = \dot{\mathbf{x}}.\delta\mathbf{x} + \dot{\mathbf{x}}.(\mathbf{r} \times \delta\alpha) + (\mathbf{r} \times \dot{\alpha}) \, \delta\mathbf{x} + (\mathbf{r} \times \dot{\alpha}) (\mathbf{r} \times \delta\alpha) .$$

Taking into account that $\int_c \mathbf{r} \, dm = 0$, it follows

$$\Pi_\delta = m\dot{\mathbf{x}} \, \delta\mathbf{x} + \theta . \delta\alpha$$

and putting

$$m = \int_c dm$$

we obtain immediately

$$\Omega_\delta^{(i)} = m\dot{\mathbf{x}}\delta\mathbf{x} + \theta.\delta\alpha ,$$

where θ is the vector $\int_c [(\mathbf{r} \times \dot{\alpha}) \times \mathbf{r}]dm$.

θ may be computed by referring $\mathbf{r}$ to a co-ordinate system associated with the body.

CHAPTER III

INVARIANTIVE MECHANICS OF SYSTEMS OF MATERIAL POINTS

§1. Inertia, Gravity and Expansion

In the case of a material point, inertia is characterized by the interdependency of mass and velocity and by the expression mc^2 of the point's energy.

In a system of several masses present in the space-time S_N, and in the absence of a field, hence in the case of inertial motion, mechanical analysis evidences the gravitational interaction and a second form of interaction that was manifested for the first time in the Hubble effect.

§2. System of n Material Points

1. Characteristic Magnitudes. Let P_j ($j = 1,2,\ldots,n$) be n material points, $\mathbf{x}_j$ and m_j their position vectors and masses respectively ; we denote by the vectors $\mathbf{p}_j$ ($j = 1,2,\ldots,n$) their impulses and by H their global energy.

We know that $m_j = m_j^0/\sqrt{1-v_j^2/c^2}$, where $v_j = \dot{x}_j$, but the impulses p_j and the energy H must be determined taking into account the following principles, which are the simple extension to the case $n > 1$ of the principles used in the case $n = 1$.

2. Principles. 1st. H is a Euclidean invariant of the geometrical system consisting of the vectors $\mathbf{p}_j$ ($j = 1,2,\ldots, n$) and the vectors $\mathbf{r}_{jk}$ of the relative positions,

$$\mathbf{r}_{jk} = \mathbf{x}_k - \mathbf{x}_j \ .$$

This means that H must be a function of the Euclidean invariants of the respective geometrical configuration.

The invariants to be considered are as follows:

$$\alpha_j = 1/2\ \mathbf{p}_j^2\ ,\quad \beta_{ij} = 1/2\ \mathbf{r}_{ij}^2$$

$$\gamma_{ij} = \mathbf{p}_i \cdot \mathbf{p}_j\ ;\quad \gamma_{ijh} = \mathbf{p}_i \cdot \mathbf{r}_{jh}$$

$$\gamma_{ik,jh} = \mathbf{r}_{ik} \cdot \mathbf{r}_{jh}\ ;$$

$$(i,j,k,h = 1, 2, \ldots, n),$$

the other invariants – for $n > 2$ – are expressed with their help.

Therefore we have

$$H = H\,(\alpha_i, \beta_{ij}, \gamma_{ij}, \gamma_{ijh}, \gamma_{ik,jh}),$$

where $i,j,h,k = 1,2, \dots n$.

2nd. The expressions of the impulses p_j and of H must satisfy the law of motion, which require that the (invariant) geometrical derivation of the system $(p_1, p_2, \dots, p_n, H)$ should vanish.

3rd. The components of the geometrical derivative of the system $(p_1, p_2, \dots, p_n, H)$ are, by definition, the coefficients of the variations $\delta x_1, \dots, \delta x_n, \delta p_1, \dots, \delta p_n, \delta t$ in the external derivative (after Cartan) of the form

$$\omega_\delta^{(i)} = \sum p_j \delta x_j - H\delta t,$$

i.e. of

$$D\omega_\delta^{(i)} = d\omega_\delta^{(i)} - \delta\omega_d^{(i)} = \sum dp_j \delta x_j - \sum dx_j \delta p_j - dH\delta t + \delta H dt\ .$$

4th. The expression of H and of the impulses must reduce to the expressions of Newton's mechanics for bodies whose masses, distances and velocities are of the usual order of magnitude, i.e. a Newtonian order of magnitude.

§3. Inertial Mechanics of a Two Body System

1. Euclidean Invariants of the System. We shall consider first the case $n = 2$, which is sufficiently complex and rich in new mechanical aspects. A simpler notation has been adopted for the invariants, since we have only one distance vector

$$\mathbf{r} = \mathbf{x}_2 - \mathbf{x}_1 .$$

We put

$$\begin{gathered} \alpha_1 = 1/2\, \mathbf{p}_1^2 , \quad \alpha_2 = 1/2\, \mathbf{p}_2^2 , \quad \alpha = \mathbf{p}_1 \cdot \mathbf{p}_2 , \\ \beta_1 = \mathbf{r} \cdot \mathbf{p}_1 , \quad \beta_2 = \mathbf{r} \cdot \mathbf{p}_2 , \quad \beta = 1/2\, \mathbf{r}^2 . \end{gathered} \tag{1}$$

Then

$$H = H(\alpha_1, \alpha_2, \alpha, \beta_1, \beta_2, \beta) .$$

We have, in that case,

$$\omega_\delta^{(i)} = \sum \mathbf{p}_j . \delta \mathbf{x}_j - H \delta t ,$$

hence

$$D\omega_\delta^{(i)} = \sum d\mathbf{p}_j \, \delta \mathbf{x}_j - \sum d\mathbf{x}_j \, \delta \mathbf{p}_j - dH \delta t + \delta H \, dt = 0 \tag{2}$$

for any δ.

In the above expression we have

$$\begin{aligned} \delta H = {} & H_{\alpha_1} \mathbf{p}_1 \, \delta \mathbf{p}_1 + H_{\alpha_2} \mathbf{p}_2 \, \delta \mathbf{p}_2 + H_\alpha \mathbf{p}_2 \, \delta \mathbf{p}_1 + H_\alpha \mathbf{p}_1 \, \delta \mathbf{p}_2 + \\ & + H_\beta \mathbf{r} \, \delta \mathbf{r} + H_{\beta_1} \mathbf{r} \, \delta \mathbf{p}_1 + H_{\beta_2} \mathbf{r} \, \delta \mathbf{p}_2 + H_{\beta_1} \mathbf{p}_1 \, \delta \mathbf{r} + H_{\beta_2} \mathbf{p}_2 \, \delta \mathbf{r} , \end{aligned} \tag{3}$$

where we have written H_α instead of $dH/d\alpha$ etc. and we must consider

$$\delta \mathbf{r} = \delta \mathbf{x}_2 - \delta \mathbf{x}_1 .$$

Then the second member of (3) becomes

$$\left(\frac{dp_1}{dt} - H_{\beta_1} p_1 - H_{\beta_2} p_2 - H_\beta r\right)\delta x_1 + \left(\frac{dp_2}{dt} + H_{\beta_1} p_1 + H_{\beta_2} p_2 + H_\beta r\right)\delta x_2 +$$

$$+ \left(-\frac{dx_1}{dt} + H_{\alpha_1} p_1 + H_\alpha p_2 + H_{\beta_1} r\right)\delta p_1 + \left(-\frac{dx_2}{dt} + H_{\alpha_2} p_2 + H_\alpha p_1 + H_{\beta_2} r\right)\delta p_2 -$$

$$(4) \qquad - \frac{dH}{dt}\,\delta t = 0 .$$

The equations of motion. By cancelling the coefficients of $\delta t, \delta x_1, \delta x_2, \delta p_1, \delta p_2$ we obtain, first, the equations of motion

$$(5) \qquad \begin{aligned} \frac{dp_1}{dt} &= H_{\beta_1} p_1 + H_{\beta_2} p_2 + H_\beta r , \\ \frac{dp_2}{dt} &= - H_{\beta_1} p_1 - H_{\beta_2} p_2 - H_\beta r , \\ \frac{dH}{dt} &= 0 \end{aligned}$$

and then the relations which will serve to define the impulses,

$$(6) \qquad \begin{aligned} v_1 &= H_{\alpha_1} p_1 + H_\alpha p_2 + H_{\beta_1} r, \\ v_2 &= H_\alpha p_1 + H_{\alpha_2} p_2 + H_{\beta_2} r, \end{aligned}$$

where

$$v_1 = \frac{dx_1}{dt} , \quad v_2 = \frac{dx_2}{dt} .$$

3. Theorems of Conservation. Equations (5) give the two theorems of conservation:

a) Conservation of the total impulse

$$(7) \qquad p_1 + p_2 = \text{const.},$$

which results by adding the first two equalities of (5).

b) Conservation of energy

$$(8) \qquad H = \text{const.},$$

resulting from the last equation of (5).

4. Determination of the Impulses p_1 and p_2. From the last equations (6) and provided that the inequality

$$\Delta = H_{\alpha_1} H_{\alpha_2} - H_\alpha^2 \neq 0 \tag{9}$$

is satisfied, we obtain the expressions

$$\begin{aligned} p_1 &= \frac{H_{\alpha_2}}{\Delta} v_1 - \frac{H_\alpha}{\Delta} v_2 + \frac{H_{\beta_1} H_\alpha - H_{\alpha_2} H_{\beta_1}}{\Delta} r, \\ p_2 &= \frac{H_{\alpha_1}}{\Delta} v_2 - \frac{H_\alpha}{\Delta} v_1 + \frac{H_{\beta_1} H_\alpha - H_{\alpha_1} H_{\beta_2}}{\Delta} r. \end{aligned} \tag{10}$$

By putting

$$m_1 = \frac{H_{\alpha_2}}{\Delta}, \quad m_2 = \frac{H_{\alpha_1}}{\Delta}, \quad \mu = -\frac{H_\alpha}{\Delta},$$

$$\ell_1 = \frac{H_{\beta_2} H_\alpha - H_{\alpha_2} H_{\beta_1}}{\Delta} = h_1 \nu, \qquad \ell_2 = \frac{H_{\beta_1} H_\alpha - H_{\alpha_1} H_{\beta_2}}{\Delta} = h_2 \nu,$$

the expressions (10) become

$$\begin{aligned} p_1 &= m_1 v_1 + \mu v_2 + h_1 \nu r, \\ p_2 &= m_2 v_2 + \mu v_1 + h_2 \nu r, \end{aligned} \tag{11}$$

where μ, which represent a mass, will be called gravity mass, and ν, to which we assign the same homogeneity of a mass, will be called by the name of Hubble.

5. Expression of the Energy H. In accordance with principle 4, the expression of H must reduce under Newtonian conditions to the form

$$H_{\text{Newton}} = 1/2\, m_1^0 v_1^2 + 1/2\, m_2^0 v_2^2 - f \frac{m_1^0 m_2^0}{r} + C. \tag{12}$$

Since H_{Newton} is definite except the constant, and since when we pass to the invariant mechanics we find for any material particle instead of $1/2\, m_1^0 v^2 + C$ the expression mc^2, we write

$$H = c^2 (m_1 + m_2 + 2\mu + 2\nu), \tag{13}$$

where the brackets include besides the individual masses also the interaction masses required by the expressions (12).

In this form H must satisfy the conditions required by the principle 2.

Since

$$m_j = \sqrt{m_j^{0\,2} + m_j^2 v_j^2/c^2},$$

we have

$$\delta(c^2 m_j) = \mathbf{v}_j . \delta(m_j \mathbf{v}_j)\ ; \quad (j = 1, 2)\ . \tag{14}$$

By (12) we have then

$$\begin{aligned} \delta(c^2 m_1) &= \mathbf{v}_1 . \delta(\mathbf{p}_1 - \mu \mathbf{v}_2 - \ell_1 \mathbf{r}) \\ \delta(c^2 m_2) &= \mathbf{v}_2 . \delta(\mathbf{p}_2 - \mu \mathbf{v}_1 - \ell_2 \mathbf{r})\ . \end{aligned} \tag{15}$$

We put for each function F of the preceding arguments

$$\delta F = \delta_1 F + \delta_2 F,$$

where $\delta_1 F$ is the variation of F resulting from a variation of r, and $\delta_2 F$ is the variation resulting from the other arguments.

Thus we have

$$\delta F = F_r \frac{\mathbf{r}.\delta \mathbf{r}}{r} + \delta_2 F\ .$$

Following this convention we obtain from (15)

$$\begin{aligned} \delta(c^2 m_1) &= \mathbf{v}_1 . \delta \mathbf{p}_1 - \mathbf{v}_1 . \mathbf{v}_2\, \delta\mu - \mu \mathbf{v}_1 . \delta \mathbf{v}_2 - \mathbf{v}_1 . \mathbf{r}\, \delta \ell_1 - \ell_1 \mathbf{v}_1 (\delta \mathbf{x}_2 - \delta \mathbf{x}_1) = \\ &= \mathbf{v}_1 . \delta \mathbf{p}_1 - \mathbf{v}_1 . \mathbf{v}_2\, \mu_r \frac{\mathbf{r}}{r} (\delta \mathbf{x}_2 - \delta \mathbf{x}_1) - \mathbf{v}_1 . \mathbf{v}_2\, \delta_2 \mu - \mu \mathbf{v}_1 . \delta \mathbf{v}_2 - \\ &\quad - \mathbf{v}_1 . \mathbf{r}\, \ell_{1,r} \frac{\mathbf{r}}{r} (\delta \mathbf{x}_2 - \delta \mathbf{x}_1) - \mathbf{v}_1\, \mathbf{r}\, \delta_2 \ell_1 - \ell_1 \mathbf{v}_1 . (\delta \mathbf{x}_2 - \delta \mathbf{x}_1)\ ; \\ \delta(c^2 m_2) &= \mathbf{v}_2 . \delta \mathbf{p}_2 - \mathbf{v}_1 . \mathbf{v}_2\, \mu_r \frac{\mathbf{r}}{r} (\delta \mathbf{x}_2 - \delta \mathbf{x}_1) - \mathbf{v}_1 . \mathbf{v}_2 . \delta_2 \mu - \mu \mathbf{v}_2 . \delta \mathbf{v}_1 - \\ &\quad - \mathbf{v}_2 . \mathbf{r}\, \ell_{2,r} \frac{\mathbf{r}}{r} (\delta \mathbf{x}_2 - \delta \mathbf{x}_1) - \mathbf{v}_2 . \mathbf{r}\, \delta_2 \ell_2 - \ell_2 \mathbf{v}_2 . (\delta \mathbf{x}_2 - \delta \mathbf{x}_1); \end{aligned} \tag{16}$$

$$\delta(c^2 \mu) = c^2 \mu_r \frac{\mathbf{r}}{r} (\delta \mathbf{x}_2 - \delta \mathbf{x}_1) + c^2 \delta_2 \mu ;$$

$$\delta(c^2 \nu) = c^2 \nu_r \frac{\mathbf{r}}{2} (\delta \mathbf{x}_2 - \delta \mathbf{x}_1) + c^2 \delta_2 \nu .$$

Hence, for H defined by (13), we obtain

$$\delta H = \mathbf{v}_1 \cdot \delta \mathbf{p}_1 + \mathbf{v}_2 \cdot \delta \mathbf{p}_2 + \big[(2\mathbf{v}_1 \cdot \mathbf{v}_2 + \ell_{1,r} \mathbf{v}_1 \cdot \mathbf{r} + \ell_{2,r} \mathbf{v}_2 \cdot \mathbf{r} -$$
$$- 2c^2 \mu_r - 2c^2 \nu_r \frac{\mathbf{r}}{r} + \ell_1 \mathbf{v}_1 + \ell_2 \mathbf{v}_2)\big](\delta \mathbf{x}_2 - \delta \mathbf{x}_1) + (2c^2 - 2\mathbf{v}_1 \cdot \mathbf{v}_2)\delta_2 \mu -$$
$$- \mu \delta(\mathbf{v}_1 \cdot \mathbf{v}_2) - \mathbf{v}_1 \cdot \mathbf{r}\, \delta_2 \ell_1 - \mathbf{v}_2 \cdot \mathbf{r}\, \delta_2 \ell_2 + 2c^2 \delta_2 \nu . \qquad (17)$$

Then from the 2nd principle and using (17) we obtain the equality

$$\left(\frac{d\mathbf{p}_1}{dt} - \mathbf{L}\right) \delta \mathbf{x}_1 + \left(\frac{d\mathbf{p}_2}{dt} + \mathbf{L}\right) \delta \mathbf{x}_2 + 2c^2 (1 - \mathbf{v}_1 \cdot \mathbf{v}_2 / c^2) \delta_2 \mu -$$
$$- \mu\, \delta(\mathbf{v}_1 \cdot \mathbf{v}_2) + 2c^2 \delta_2 \nu - \mathbf{v}_1 \cdot \mathbf{r} . \delta_2 \ell_1 - \mathbf{v}_2 \cdot \mathbf{r}\, \delta_2 \ell_2 - \frac{dH}{dt} \delta t = 0, \qquad (18)$$

where

$$\mathbf{L} = \big[2c^2 (1 - \mathbf{v}_1 \cdot \mathbf{v}_2 / c^2) \mu_r + 2c^2 \nu_r - \mathbf{v}_1 \cdot \mathbf{r}\, \ell_{1,r} - \mathbf{v}_2 \cdot \mathbf{r}\, \ell_{2,r}\big] \frac{\mathbf{r}}{r} - \ell_1 \mathbf{v}_1 - \ell_r \mathbf{v}_2 . \qquad (19)$$

Hence we obtain the equations of motion in the form

$$\frac{d\mathbf{p}_1}{dt} = \mathbf{L} , \quad \frac{d\mathbf{p}_2}{dt} = - \mathbf{L} , \quad H = H_0 . \qquad (20)$$

The second members of the first two equations correspond in Newtonian language to the forces, equal and of opposite direction, which the two bodies exert on each other. The constant energy H is the sum of the two energies of the bodies P_1 and P_2, of the gravitational energy $2\, c^2 \mu$ and of an interaction energy $2\, c^2 \nu$, the structure of which will be explained below.

The law $H = H_0$ corresponds for an inertial system to the conservation of the total mass

$$m_1 + m_2 + 2\mu + 2\nu = m_1^0 + m_2^0 + 2\mu^0 + 2\nu^0 ,$$

consisting of the individual masses, the mass of gravitational interaction (or Newton's mass) and of the interaction mass ν.

6. Determination of the mass μ of gravitational interaction. Separating in relation (18) the terms containing μ, we obtain

$$2c^2(1-\mathbf{v}_1\cdot\mathbf{v}_2/c^2) = \mu\delta_2(\mathbf{v}_1\cdot\mathbf{v}_2),$$

where, in the second member, we have δ_2 instead of δ, since r does not appear directly in the brackets.

Integrating, we obtain

$$\mu = \frac{\varphi(r)}{\sqrt{1-\mathbf{v}_1\cdot\mathbf{v}_2/c^2}}, \tag{21}$$

where $\varphi(r)$ is till now an arbitrary function of r.

Comparing with (12) it is clear that we must take

$$\varphi(r) = -1/2\, f\,\frac{m_1^0 m_2^0}{c^2 r} + \frac{K}{c^4 r^2}. \tag{22}$$

where f is the Newtonian constant.

But the form of the function φ permits also to take into account the observed motion of the perihelion (see the note by I. Mihaila, page 118). Consequently it follows that

$$\mu = -\frac{f}{2c^2}\,\frac{m_1^0\, m_2^0}{r\sqrt{1-\mathbf{v}_1\cdot\mathbf{v}_2/c^2}}\,(1+\frac{\lambda(r)}{c^2}) \tag{23}$$

where $\lambda(r) = k/r$, $k = 9/2\, f(m_1^0 + m_2^0)$.

In the case of usual velocities, μ is practically equal to $-1/2\, f m_1^0 . m_2^0/c^2 r$ and the energy is that which appears in the Newtonian theory, i.e. $-f m_1^0\, m_2^0/r$.

Terrestrial and ordinary astronomical experience extended to the solar system and even to celestial bodies pertaining to our galaxy, does not require the consideration of elements other than the preceding ones; therefore the mass ν may be taken to be zero.

For such systems we shall have then, with a sufficient approximation

$$p_1 = m_1 v_1 + \mu v_2,$$

$$p_2 = m_2 v_2 + \mu v_2,$$

$$m_1 = m_1^0 / \sqrt{1 - v_1^2/c^2}, \quad m_2 = m_2^0 / \sqrt{1 - v_2^2/c^2},$$

$$\mu = - \frac{f}{2c^2} \frac{m_1^0 m_1^0}{r\sqrt{1 - v_1 \cdot v_2 / c^2}} \left(1 + \frac{k}{c^2 r}\right),$$

$$H = c^2 (m_1 + m_2 + 2\mu),$$

and the equations of motion are

$$\frac{dp_1}{dt} = L, \quad \frac{dp_2}{dt} = - L,$$

where

$$L = - 2c^2 (1 - v_1 \cdot v_2 / c^2) \frac{r}{r} \mu_r .$$

Note: When the distance r is of a nuclear order of magnitude, the above equations take a special form, in which account is taken of the small magnitude of r and of the considerably smaller magnitudes of m_1^0 and m_2^0.

7. Determination of the interaction mass ν. After determining μ, condition (18) gives the last relation

$$2c^2 \delta_2 \nu - v_1 \cdot r \cdot \delta_2 \ell_1 - v_2 r \delta_2 \ell_2 = 0, \tag{24}$$

where the mass ν is included in the previously indicated expressions of ℓ_1 and ℓ_2:

$$\ell_1 = h_1 \nu, \quad \ell_2 = h_2 \nu .$$

Hence

$$\delta_2 \ell_1 = h_1 \delta_2 \nu + \nu \delta_2 h_1 ; \quad \delta_2 \ell_2 = h_2 \delta_2 \nu + \nu \delta_2 h_2 . \tag{25}$$

Then relation (24) becomes

$$(2c^2 - h_1 v_1 . r - h_2 v_2 . r) \delta_2 \nu = (v_1 . r \delta_2 h_1 + v_2 . r \delta_2 h_2) \nu, \tag{26}$$

whence

$$\left(1-\frac{h_1 \mathbf{v}_1.\mathbf{r}+h_2 \mathbf{v}_2.\mathbf{r}}{2c^2}\right)\frac{\delta_2 \nu}{\nu}=\frac{\mathbf{v}_1.\mathbf{r}\delta_2 h_1+\mathbf{v}_2.\mathbf{r}.\delta_2 h_2}{2c^2} .$$

We seek expressions of the functions h_1 and h_2 for which the preceding equation admits a solution of the form

$$\nu = \psi(\mathbf{r})/\sqrt{1-\frac{h_1 \mathbf{v}_1.\mathbf{r}+h_2 \mathbf{v}_2 \mathbf{r}}{2c^2}} , \tag{27}$$

indicated by (26), and we obtain

$$h_1 = \psi_1(\mathbf{r})\,\mathbf{v}_1.\mathbf{r} , \quad h_2 = \psi_2(\mathbf{r})\,\mathbf{v}_2.\mathbf{r},$$

which gives for the expression under the radical

$$u=\frac{h_1 \mathbf{v}_1.\mathbf{r}+h_2 \mathbf{v}_2 \mathbf{r}}{2c^2}=\frac{\psi_1(\mathbf{r})(\mathbf{v}_1.\mathbf{r})^2+\psi_2(\mathbf{r})(\mathbf{v}_2.\mathbf{r})^2}{2c^2} ; \tag{28}$$

it follows that, in order to obtain a real value of ν, for any r, we must take

$$\psi_1(\mathbf{r}) = \frac{m_1^0}{m_1^0+m_2^0}\frac{1}{r^2} , \quad \psi_2(\mathbf{r}) = \frac{m_2^0}{m_1^0+m_2^0}\frac{1}{r^2} . \tag{29}$$

Hence

$$\nu = \psi(\mathbf{r})/\sqrt{1-u} , \quad u=\frac{m_1^0 v_1^2 \cos^2(\mathbf{v}_1,\mathbf{r})+m_2^0 v_2^2 \cos^2(\mathbf{v}_2,\mathbf{r})}{2(m_1^0+m_2^0)c^2} , \tag{30}$$

where $\psi(\mathrm{r})$ is not yet determined. We remark that in the case of subgalactic distances, ν must practically vanish. Its values become significant only when r takes values whose order of magnitude is at least c^2.

For this reason we put

$$\psi(\mathbf{r}) = -\frac{1}{2}g\left(\frac{r}{c^2}\right)^2 , \tag{31}$$

where g can depend on m_1^0, m_2^0.

This gives for the invariant mechanics the expressions

$$H = c^2 m_1 + c^2 m_2 - f \frac{m_1^0 m_2^0}{r\sqrt{1 - \mathbf{v}_1 \cdot \mathbf{v}_2 / c^2}} \left(1 + \frac{k}{c^2 r}\right) - g \frac{r^2}{c^2 \sqrt{1-u}} ,$$

$$p_1 = m_1 \mathbf{v}_1 - \frac{1}{2} \frac{f m_1^0 m_2^0 \mathbf{v}_2}{c^2 r \sqrt{1 - \mathbf{v}_1 \cdot \mathbf{v}_2 / c^2}} \left(1 + \frac{k}{c^2 r}\right) - \frac{1}{2} g \frac{r m_1^0 v_1 \cos(\mathbf{v}_1, \mathbf{r})}{(m_1^0 + m_2^0) c^4 \sqrt{1-u}} \mathbf{r} ,$$

$$p_2 = m_2 \mathbf{v}_2 - \frac{1}{2} \frac{f m_1^0 m_2^0 \mathbf{v}_1}{c^2 r \sqrt{1 - \mathbf{v}_1 \cdot \mathbf{v}_2 / c^2}} \left(1 + \frac{k}{c^2 r}\right) - \frac{1}{2} g \frac{r m_2^0 v_2 \cos(\mathbf{v}_2 \cdot \mathbf{r})}{(m_1^0 + m_2^0) c^4 \sqrt{1-u}} \mathbf{r} . \tag{32}$$

§ 4. The expansion of the System and Hubble's Law

1. Hubble's Law. Hubble's empirical law

$$|v_2 \cos(v_2, r) - v_1 \cos(v_1, r)| = \theta r,$$

where θ is a constant, is valid as a law of mechanics for supergalactic distances at least for some systems and in an adequate time interval.

2. Law of Expansion. From (7) and (11), we obtain

$$m_1 v_1 + m_2 v_2 + \mu(v_1 + v_2) + (h_1 + h_2)\nu r = C. \tag{33}$$

Following a scalar multiplication by r, considering the expressions of h_1 and h_2 and remarking that $v_j r = v_j r \cos(v_j, r) = r\xi_j$, $\xi_j = v_j \cos(v_j, r)$, relation (33) becomes, after division by r,

$$m_1\xi_1 + m_2\xi_2 + \mu(\xi_1 + \xi_2) + \frac{m_1^0\xi_1 + m_2^0\xi_2}{m_1^0 + m_2^0} \frac{1}{r^2}\nu r^2 = C\cos(C,r) = \Gamma .$$

Dividing again by $m_1\xi_j + m_2\xi_2$, provided that the latter expression is different from zero, and replacing μ and ν by their expressions (23) and (30), we obtain the equation

$$1 + \frac{1}{c^2 r} M' + \frac{1}{c^4 r} M'' + \frac{1}{c^4} N' r^2 = \frac{\Gamma}{m_1\xi_1 + m_2\xi_2}, \tag{34}$$

where

$$M' = -\frac{1}{2} f \frac{m_1^0 m_2^0}{\sqrt{1 - \upsilon_1 . \upsilon_2 \epsilon c^2}} \cdot \frac{\xi_1 + \xi_2}{m_1\xi_1 + m_2\xi_2};$$

$$M'' = kM'$$

$$N' = -\frac{1}{2} g \frac{m_1^0 m_2^0}{m_1^0 + m_2^0} \frac{m_1^0\xi_1 + m_2^0\xi_2}{m_1\xi_1 + m_2\xi_2} \cdot \frac{1}{\sqrt{1-u}}$$

Putting then

$$M_1 = M / \left(\frac{\Gamma}{m_1\xi_1 + m_2\xi_2} - 1\right), \quad M_2 = M''/\left(\frac{\Gamma}{m_1\xi_1 + m_2\xi_2} - 1\right), \tag{35}$$

$$N = N'/\left(\frac{\Gamma}{m_1\xi_1 + m_2\xi_2} - 1\right),$$

we obtain the equation

$$\frac{1}{c^2 r} M_1 + \frac{1}{c^4 r^2} M_2 + \frac{r^2}{c^4} N = 1 . \qquad (36)$$

By differentiating (36) with respect to t and dividing by $2\, r/c^4 N$, we obtain he equation

$$\frac{c^2}{2r^2}\frac{\dot{M}_1}{N} + \frac{1}{2r^3} M_2 - \frac{c^2}{2r^3}\frac{M_1}{N}\frac{\mathbf{r}.\dot{\mathbf{r}}}{r} + \frac{\mathbf{r}.\dot{\mathbf{r}}}{r^4}\frac{M_2}{N} + \frac{\mathbf{r}.\dot{\mathbf{r}}}{r} + \frac{1}{2}\frac{\dot{N}}{N} r = 0$$

that is

$$\left(1 - \frac{c^2}{2r^3}\frac{M_1}{N} - \frac{1}{r^3}\frac{M_2}{N}\right)\frac{\mathbf{r}.\dot{\mathbf{r}}}{r} = -\frac{1}{2}\frac{\dot{N}}{N}\left(1 + \frac{c^2}{r^3}\frac{\dot{M}_1}{\dot{N}} + \frac{1}{r^4}\frac{\dot{M}_2}{\dot{N}}\right). \qquad (37)$$

Since

$$\frac{\mathbf{r}.\dot{\mathbf{r}}}{r} = v_2 \cos(v_2, \mathbf{r}) - v_1 \cos(v_1, \mathbf{r}) = W$$

represent the velocity of recession of the two bodies, equation (37) takes the form

$$|W| = \frac{1}{2}\left|\frac{\dot{N}}{N}\right| \left| \frac{1 + \frac{c^2}{r^3}\frac{\dot{M}_1}{\dot{N}} + \frac{1}{r^4}\frac{\dot{M}_2}{\dot{N}}}{1 - \frac{c^2}{2r^3}\frac{M_1}{N} - \frac{1}{r^3}\frac{M_2}{N}} \right| r, \qquad (38)$$

which constitutes an exact law of Mechanics.

The case of the galaxies. By (35) we may have $m_1 \xi_1 + m_2 \xi_2 = 0$ or $m_1^0 \xi_1 + m_2^0 \xi_2 = ($ only in very special cases, which shall be excluded; we shall consider only systems for which there exists a $k(0 < k < 1)$, such that

$$kc(m_1 + m_2) < |m_1 \xi_1 + m_2 \xi_2| < c(m_1 + m_2) ,$$

$$kc(m_1^0 + m_2^0) < |m_1^0 \xi_1 + m_2^0 \xi_2| < c(m_1^0 + m_2^0).$$

We assume also that $v_j \leqslant 9/10\, c (j = 1,2)$, and by immediate evaluations we find

$$\frac{M}{N} < \frac{f}{g} \cdot \frac{1}{k} \cdot$$

Furthermore, we see that there exists a function $U(0 \leqslant U \leqslant 1)$ of those elements of the motion, which occur directly in M and N and we consider the systems for which $|U| \geqslant U_0$ for any arbitrarily small U_0. Then we obtain

$$\frac{N}{N} = U\alpha ,$$

where

$$\alpha = |\dot{U}| + \frac{5}{k}\,(|\dot{\xi}_1| + |\dot{\xi}_2|)$$

and

$$\left|\frac{\dot{M}}{\dot{N}}\right| = \left|\frac{\dot{M}}{NU\alpha}\right| = \frac{f}{g}\,\frac{|V|}{|U|}\;\frac{\ell_1 |v_1| + \ell_2 |v_2| + |\xi_1| + |\xi_2|}{|\dot{U}| + \frac{5}{k}\,(|\dot{\xi}_1| + |\dot{\xi}_2|)} \cdot$$

By making the assumption, quite adequate for the motions considered, that the accelerations $\dot{v}_1$, $\dot{v}_2$, $\dot{\xi}_1$, $\dot{\xi}_2$ must be within two sufficiently close and small limits a_1 and a_2, we obtain

$$\left|\frac{\dot{M}}{\dot{N}}\right| < \frac{f}{gU_0} \cdot \frac{(\ell_1 + \ell_2 + 2)a_2}{(1 + \frac{5}{k})\,a_1} \cdot$$

Therefore, even for distances such as $r = lc(l > 1)$, the exact law of mechanics may be replaced by the approximate law

$$|W| \approx \frac{1}{2}\left|\frac{\dot{N}}{N}\right| r \; . \tag{39}$$

This shows that the velocity of recession of the two galaxies considered is proportional to r. The coefficient $1/2\,|\dot{N}/N|$ remains within very close limits with very slow variations.

It follows, that the law (10) which corresponds to Hubble's empirical law, is a new form whereby the inertia of matter becomes manifest.

Thus the unity of matter throughout the whole universe becomes apparent by the coexistence of gravity and expansion.

CHAPTER IV

INERTIAL MOVEMENT OF A SYSTEM OF TWO PARTICLES

§1. Definition of the Stable Particles

A body that conserves in certain limits – during its movement – the characteristic of the rigid body of the classical mechanics was defined [Ch.II] as follows:

The dynamic mass has the expression

$$(1)\qquad m = \sqrt{m_0^2 + \frac{2}{c^2}\Big(\alpha + \sum_{j=1,2,3} u_j\,\omega_j + \sum_{j=1,2,3} v_j\,\varphi_j\Big)},$$

where $\alpha = 1/2\, mv^2$, $\omega_j = 1/2\theta_j^2$, and the φ_j are nonholonomic quantities defined by the following relations

$$(2)\qquad \delta\varphi_1 = \theta_2\,\delta\theta_3 - \theta_3\,\delta\theta_2\,;\quad \delta\varphi_2 = \theta_3\,\delta\theta_1 - \theta_1\,\delta\theta_3\,;\quad \delta\varphi_3 = \theta_1\,\delta\theta_2 - \theta_2\,\delta\theta_1$$

for which we have

$$(3)\qquad \sum_{j=1,2,3} u_j\,\delta\omega_j + \sum_{j=1,2,3} \nu_j\,\delta\varphi_j = m \sum_{j=1,2,3} \dot{\alpha}_j\,\delta\theta_j\,,$$

α_1, α_2, α_3 being the angles of the orientation of the body and θ_1, θ_2, θ_3 the respective impulsions, defined by the equalities

$$(4)\qquad \begin{aligned} \theta_1 &= (\ell_{22} + \ell_{33})\dot{\alpha}_1 - \ell_{12}\,\dot{\alpha}_2 + \ell_{13}\,\dot{\alpha}_3\,,\\ \theta_2 &= \ell_{21}\,\dot{\alpha}_1 + (\ell_{33} + \ell_{11})\dot{\alpha}_2 - \ell_{23}\,\dot{\alpha}_3\,,\\ \theta_3 &= -\,\ell_{31}\,\dot{\alpha}_1 + \ell_{32}\,\dot{\alpha}_2 + (\ell_{11} + \ell_{22})\dot{\alpha}_3\,, \end{aligned}$$

where

$$(5)\qquad \ell_{jk} = \int_c \zeta_j\,\zeta_k\,dm\,,$$

$\zeta_1, \zeta_2, \zeta_3$ being the components of the vector $\boldsymbol{\rho} - \mathbf{x}$, with $\boldsymbol{\rho}$ the position vector of a generic point of the body and $\mathbf{x}$ the position vector of the mass center. A supplementary condition is that the expressions

$$\frac{1}{m} \ell_{jk}$$

are practically constant during the movement; this condition, implies, generally a limitation of the extension of the body.

The energy of each stable particle is given by the relation

$$H = mc^2 \tag{6}$$

as for a material point.

§2. The Inertial System of Two (Stable) Particles

1. Let C_1 and C_2 be a system of particles of the preceding type, $\mathbf{x}_1$ and $\mathbf{x}_2$ being their mass centers, α_1 and α_2 the respective angles of rotation. The impulsions corresponding to the positions and to the angles will be $\mathbf{p}_i$ and θ_i $(i = 1,2)$.

Following the principle of the invariantive mechanics we must first make precise the Euclidean invariants of the dynamic state of the system.

They are, evidently

$$(7)\quad \alpha_1 = \frac{1}{2}\mathbf{p}_1^2,\ \alpha_2 = \frac{1}{2}\mathbf{p}_2^2,\ \alpha = \mathbf{p}_1\cdot\mathbf{p}_2,\ \beta_1 = \mathbf{p}_1\cdot\mathbf{r},\ \beta_2 = \mathbf{p}_2\cdot\mathbf{r},\ \beta = \frac{1}{2}\mathbf{r}^2,$$
$$\gamma_1 = \frac{1}{2}\theta_1^2,\ \gamma_2 = \frac{1}{2}\theta_2^2,\ \gamma = \theta_1\cdot\theta_2,\ \epsilon_1 = \theta_1\cdot\mathbf{r},\ \epsilon_2 = \theta_2\,\mathbf{r},$$

with $\mathbf{r} = \mathbf{x}_2 - \mathbf{x}_1$.

The energy H of the system will be an invariant of the system, then

$$H = H(\alpha_1, \alpha_2, \alpha, \beta_1, \beta_2, \beta, \gamma_1, \gamma_2, \gamma, \epsilon_1, \epsilon_2),$$

from which it results, putting $\delta\mathbf{r} = \delta\mathbf{x}_2 - \delta\mathbf{x}_1$,

$$(8)\quad \begin{aligned} \delta H = {} & H_{\alpha_1}\mathbf{p}_1\delta\mathbf{p}_1 + H_{\alpha_2}\mathbf{p}_2\delta\mathbf{p}_2 + H_\alpha\mathbf{p}_2\delta\mathbf{p}_1 + H_\alpha\mathbf{p}_1\delta\mathbf{p}_2 + H_{\beta_1}\mathbf{r}\delta\mathbf{p}_1 + \\ & + H_{\beta_1}\mathbf{p}_1(\delta\mathbf{x}_2 - \delta\mathbf{x}_1) + H_{\beta_2}\mathbf{r}\delta\mathbf{p}_2 + H_{\beta_2}\mathbf{p}_2(\delta\mathbf{x}_2 - \delta\mathbf{x}_1) + H_{\gamma_1}\theta_1\,\delta\theta_1 + \\ & + H_{\gamma_2}\theta_2\cdot\delta\theta_2 + H_\gamma\theta_2\,\delta\theta_1 + H_\gamma\theta_1\cdot\delta\theta_2 + H_\beta\,\mathbf{r}(\delta\mathbf{x}_2 - \delta\mathbf{x}_1) + H_{\epsilon_1}\mathbf{r}\cdot\delta\theta_1 + \\ & + H_{\epsilon_1}\theta_1(\delta\mathbf{x}_2 - \delta\mathbf{x}_1) + H_{\epsilon_2}\mathbf{r}\cdot\delta\theta_2 + H_{\epsilon_2}\theta_2(\delta\mathbf{x}_2 - \delta\mathbf{x}_1). \end{aligned}$$

The inertial fundamental form of the system will be

$$(9)\quad \omega_\delta^{(i)} = \mathbf{p}_1\,\delta\mathbf{x}_1 + \mathbf{p}_2\,\delta\mathbf{x}_2 + \theta_1\,\delta\alpha_1 + \theta_2\,\delta\alpha_2 - H\delta t\,.$$

If we cancel the coefficients of $\delta\mathbf{x}_1$, $\delta\mathbf{x}_2$ in the Cartan (or exterior) derivative of $\omega_\delta^{(i)}$, i.e. in the expression of $d\omega_\delta^{(i)} - \delta\omega_d^{(i)}$, we obtain, after division by dt, the equations

$$\frac{dp_1}{dt} = -M, \quad \frac{dp_2}{dt} = M, \tag{10}$$

where

$$M = H_\beta r + H_{\beta_1} p_1 + H_{\beta_2} p_2 + H_{\epsilon_1} \theta_1 + H_{\epsilon_2} \theta_2 , \tag{11}$$

that gives immediately the prime integral of the conservation of the total impulsion

$$p_1 + p_2 = C . \tag{12}$$

If we cancel in the derivative of $\omega_\delta^{(i)}$ the coefficients of δt, $\delta\theta_1$, $\delta\theta_2$, we obtain the laws of conservation of the energy

$$H = H_o \tag{13}$$

and of the angular impulsions:

$$\theta_1 = \theta_1^0 , \quad \theta_2 = \theta_2^0 . \tag{14}$$

Cancelling the coefficients of δp_1, δp_2, $\delta\theta_1$, $\delta\theta_2$, we obtain the relations binding the impulsions to the respective velocities:

$$\begin{aligned} v_1 &= H_{\alpha_1} p_1 + H_\alpha p_2 + H_{\beta_1} r , \\ v_2 &= H_{\alpha_2} p_2 + H_\alpha p_1 + H_{\beta_2} r \end{aligned} \tag{15}$$

and

$$\begin{aligned} \dot{\alpha}_1 &= H_{\gamma_1} \theta_1 + H_\gamma \theta_2 + H_{\epsilon_1} r , \\ \dot{\alpha}_2 &= H_{\gamma_2} \theta_2 + H_\gamma \theta_1 + H_{\epsilon_2} r . \end{aligned} \tag{16}$$

It will be supposed that

$$H_{\alpha_1} H_{\alpha_2} - H_\alpha^2 \neq 0 \qquad H_{\gamma_1} H_{\gamma_2} - H_\gamma^2 \neq 0$$

and consequently that the relations (15) and (16) can be inversed. First we obtain from (15) the expressions

$$(17)\qquad \begin{aligned} p_1 &= m_1 v_1 + \mu v_2 + h_1 \nu r , \\ p_2 &= m_2 v_2 + \mu v_1 + h_2 \nu r , \end{aligned}$$

similar to those of the material points: the coefficients m_1 and m_2 will be the respective dynamical masses, μ the gravitation mass, and ν the dilatational mass.

From (16) we obtain the expressions of θ_1 and θ_2 :

$$(18)\qquad \begin{aligned} \theta_1 &= m_1' \dot{\alpha}_1 + \mu' \dot{\alpha}_2 + h_1' \nu' r , \\ \theta_2 &= m_2' \dot{\alpha}_2 + \mu' \dot{\alpha}_1 + h_2' \nu' r . \end{aligned}$$

The coefficients m_1', m_2', μ' and ν' can be called angular masses, but they have not yet, for us, a physical interpretation.

2. The Determination of H. We shall put as in the case of two material points

$$(19)\qquad H = c^2 (m_1 + m_2 + 2\mu + 2\nu) ,$$

where μ is the gravitational mass and ν the dilatational mass of the system, neglecting the contributions at the energy of the angular masses m_1', m_2', μ' and ν'.

If we consider two bodies similar to the sun and a planet, the dilatational mass ν will be neglected and the expression of H becomes

$$(20)\qquad H = C^2 (m_1 + m_2 + 2\mu) .$$

We observe that, according to (1) and (2), we can put

$$\delta(c^2 m_1) = v_1 \delta(m_1 v_1) + \dot{\alpha}_1 \delta\theta_1 ,$$

$$\delta(c^2 m_2) = v_2 \delta(m_2 v_2) + \dot{\alpha}_2 \delta\theta_2 ,$$

$$\delta(c^2 \mu) = c^2 \mu_r \frac{r}{r} (\delta x_2 - \delta x_1) + c^2 \delta' \mu ,$$

where $\delta'\mu$ represent the variation of μ with respect to the other variables as r. These expressions become:

$$\delta(c^2 m_1) = \mathbf{v}_1\, \delta(\mathbf{p}_1 - \mu \mathbf{v}_2) + \dot{\alpha}_1 \delta\theta_1 = \mathbf{v}_1\, \delta\mathbf{p}_1 - \mathbf{v}_1 \cdot \mathbf{v}_2\, \mu_r \frac{\mathbf{r}}{r}(\delta\mathbf{x}_2 - \delta\mathbf{x}_1) - \\ - \mathbf{v}_1 \cdot \mathbf{v}_2\, \delta'\mu - \mu \mathbf{v}_1 . \delta\mathbf{v}_2 + \dot{\alpha}_1 \delta\theta_1 \tag{21}$$

and the similar expression for $\delta(c^2 m_2)$.

It results that

$$\delta H = \mathbf{v}_1 \delta\mathbf{p}_1 + \mathbf{v}_2 . \delta\mathbf{p}_2 - 2\mathbf{v}_1 \cdot \mathbf{v}_2\, \mu_r \frac{\mathbf{r}}{r}(\delta\mathbf{x}_2 - \delta\mathbf{x}_1) - 2\mathbf{v}_1 \cdot \mathbf{v}_2\, \delta'\mu - \\ - \mu\delta(\mathbf{v}_1 \cdot \mathbf{v}_2) + c^2 \mu_r \frac{\mathbf{r}}{r}(\delta\mathbf{x}_2 - \delta\mathbf{x}_1) + \dot{\alpha}_1 \delta\theta_1 + \dot{\alpha}_2 \delta\theta_2 + 2c^2 \delta'\mu \ . \tag{22}$$

3. <u>The equations of the movement</u> result by the cancellation of the coefficients of $\delta\mathbf{x}_1$ and $\delta\mathbf{x}_2$ in the Cartan derivative of $\omega_\delta^{(i)}$ and are

$$\frac{d\mathbf{p}_1}{dt} = \mathbf{L}\,; \quad \frac{d\mathbf{p}_2}{dt} = -\mathbf{L}\ , \tag{23}$$

where

$$\mathbf{L} = 2c^2(1 - \mathbf{v}_1 . \mathbf{v}_2 / c^2)\frac{\mathbf{r}}{r} \tag{24}$$

and

$$\mu = \frac{\varphi(r)}{\sqrt{1 - \mathbf{v}_1 . \mathbf{v}_2 / c^2}} \tag{25}$$

with

$$\varphi(r) = \frac{1}{2} f' \frac{m_1^0\, m_2^0}{c^2 r}\left(1 + \frac{k}{c^2 r}\right) . \tag{26}$$

In a first approximation, very near to the case in which the two bodies can be replaced by material points,we can take f' = f like Newton did. But otherwise, we must determine f' for each particular case.

We obtain, as in the general case

$$\theta_1 = \theta_1^0\ , \qquad \theta_2 = \theta_2^0\ . \tag{27}$$

Consequently we will have for the masses m_1 and m_2 the simplified expressions

$$m_i = \sqrt{m_i^{02} + \frac{2}{c^2}\,(1/2\, m_i^2\, v_i^2 + K_i)}\ ,$$

where K_i are also constants.

The interpretation of the relations (27) according to the fact that θ_1 and θ_2 are constant must take the object of ulterior studies.

CHAPTER V

MECHANICS OF CONTINUOUS SYSTEMS

§1. Principles

The principles which govern the characteristic relations of the motion of a continuous system are represented by the following requirements which are a continuation of the program effected in the case of a single material particle or of a finite system of material particles:

1°. The system is considered to be made up of parts, the dimensions of which are sufficiently small to assimilate each of them to a stable particle in the sense of chapter II.

2°. Establishing the relations corresponding to the pure inertial motion of each of these components. (This part of the program has been accomplished in chapter II.)

3°. Establishing the relations of motion of each component taking into account the constraints to which they are subjected inside the system.

4°. Combining the above relations into a single integral relation with the help of the coefficients corresponding to the structure of the system in the form $\omega_\delta^{(i)} + \delta S$.

5°. Establishing the elementary expression $D\omega_\delta{}^{(p)}$ corresponding to the presence of the field.

6°. Writing the conditions for the motion at the boundary imposed on the system by the external constraints. Including if possible these relations in $D\Omega_\delta{}^{(p)}$ or associating them with an additional term $\delta S^{(e)}$.

7°. Establishing the equations of motion and the equation of the energy from the relation $D(\omega_\delta{}^{(i)} + \delta S) = D(\Omega_\delta{}^{(p)} + \delta S^{(e)})$.

§2. Pure Inertial Motion of the Components

1. Let D^o be the finite and simply connected domain in E_3 taken up by the given continuous system at the time $t = 0$; let S^o be its boundary in space; let D be the domain taken up at the time t; and let S be the respective boundary. Let also (ξ, t) be the position at the time of t of a generic point of the system whose position at the time $t = 0$ was at the point ξ^o; let $\dot{\xi}$ be its velocity and $\dot{\xi}_1, \dot{\xi}_2, \dot{\xi}_3$ the velocity components.

We now consider inside of D^o a simply connected domain d^o having a regular, simple and closed surface s^o, the magnitude of which is small enough that d^o may be looked upon as a stable particle in the sense of chapter II; d and s will be what d^o and s^o become at the time t by the effect of motion.

The generating form of the relations of pure inertial motion will be, as has been shown in chapter II,

(1) $$\Omega_\delta^{(i)} = \int_d (\dot{\xi}.\delta\xi - h\,\delta t)\,dm,$$

where h is the energy per unit mass of the material particle (ξ, t) of mass dm.

Using the notations of chapter II,we obtain

$$\Omega_\delta^{(i)} = \mathbf{p}\,\delta\mathbf{x} + \sum_j \theta_j\,\delta\alpha_j - mc^2\,\delta t,$$

with

(2) $$\mathbf{p} = m\dot{\mathbf{x}}, \quad m = \int dm = \sqrt{m_0^2 + \frac{2}{c^2}\,(\alpha + \sum u_j\omega_j + \sum_j v_j\varphi_j)},$$

where α, w_j, φ_j take values determined at the same time as the motion:

$$\alpha = \frac{1}{2}\mathbf{p}^2\,; \quad \omega_j = \frac{1}{2}\theta_j^2,$$

(3) $$\delta\varphi_1 = \theta_2\,\delta\theta_3 - \theta_3\,\delta\theta_2\,,\ \delta\varphi_2 = \theta_3\,\delta\theta_1 - \theta_1\,\delta\theta_3\,,$$

$$\delta\varphi_3 = \theta_1\,\delta\theta_2 - \theta_2\,\delta\theta_1\,.$$

The coefficients $u_1, u_2, u_3, v_1, v_2, v_3$ (see chapter II) are dependent only on the spatial structure of the system.

It is clear that in the case of continuous systems of usual dimensions and for current translational and rotational velocities we have

$$m = m_0\,.$$

Under these conditions the principle $D\Omega_\delta^{(i)} = d\Omega_\delta^{(i)} - \delta\Omega_d^{(i)} = 0$ gives the relations of pure inertial motion in the form

(4) $$\rho\,\frac{d\mathbf{f}}{dt}\,\delta\mathbf{x} + \rho\sum_j \frac{d\theta_j}{dt}\,\delta\alpha_j - \rho\,\frac{dh}{dt}\,\delta t = 0,$$

where $\mathbf{f} = \mathbf{v} = d\mathbf{x}/dt$ in the case of a fluid and $\mathbf{f} = d\mathbf{u}/dt$, where $\mathbf{u}$ is the displacement vector, for elastic bodies.

§3. The Relations Imposed by the Internal Constraints

a) The principle which we shall designate as the stability of expansion requires the stability of the integral of the expression

$$(5) \qquad \mathcal{E}_d = \int_{d_0} \operatorname{div} \varphi . a(\xi, \alpha, t)\, d\tau_0 \,,$$

where the vector φ is the displacement $\mathbf{u}$ for elastic bodies, and the velocity $\mathbf{v}$ for fluids;

The multiplier $a(\xi, \alpha, t)$ is introduced in order to characterize the non-homogeneity of the medium.

The magnitudes φ, ξ and α are considered at the time t and are expressed as functions of initial values.

The condition of stability requires the equality

$$(6) \qquad \delta \mathcal{E}_d = 0$$

i.e.

$$(7) \qquad \int_{d_0} \delta(\operatorname{div} \varphi . a(\xi, \alpha, t))\, d\tau_0 = 0 .$$

But

$$\delta(\operatorname{div} \varphi\, a(\xi, \alpha, t)) = \operatorname{grad}_\xi(a \operatorname{div} \varphi)\, \delta\xi + \operatorname{grad}_\alpha (a \operatorname{div} \varphi)\, \delta\alpha + \frac{\partial}{\partial t}(a \operatorname{div} \varphi)\, \delta t$$

and $\delta\xi = \delta \mathbf{x} + \mathbf{r} \times \delta \mathbf{x}$; hence equality (7) becomes

$$(8) \qquad \mathbf{L} . \delta \mathbf{x} + \mathbf{M} . \delta\alpha + N . \delta t = 0 ,$$

where

$$(9) \qquad \mathbf{L} = \int_{d_0} \operatorname{grad}_\xi (a \operatorname{div} \varphi)\, d\tau_0 = \int_d \operatorname{grad}_\xi (a \operatorname{div} \varphi)\, \Delta\, d\tau ,$$

$$(10) \qquad \mathbf{M} = \int_d \left[(\operatorname{grad}_\xi (a \operatorname{div} \varphi) \times \mathbf{r}) + \operatorname{grad}_\alpha (a \operatorname{div} \varphi) \right] \Delta\, d\tau ,$$

$$N = \int_d \frac{\partial}{\partial t}(a \operatorname{div} \varphi)\, \Delta d\tau, \tag{11}$$

$\Delta = d\tau_0 / d\tau$ being the functional determinant of the transformation which carries the initial co-ordinates ξ_0 into the coordinates at the time t.

If we divide the first members of (3) by $\int_d \Delta . d\tau \; dt$ and if d approaches its center of mass **x**, the relation becomes

$$\operatorname{grad}_x (a \operatorname{div} \varphi)\, \delta \mathbf{x} + \operatorname{grad}_\alpha (a \operatorname{div} \varphi)\, \delta\alpha + \frac{\partial}{\partial t}(a \operatorname{div} \varphi)\, \delta t = 0 \tag{12}$$

in the regular case where $\operatorname{grad}_\xi (a \operatorname{div} \varphi) \times r \to 0$. This relation takes the form

$$\operatorname{grad}_x (a \operatorname{div} \mathbf{u})\, \delta \mathbf{x} + \operatorname{grad}_\alpha (a \operatorname{div} \mathbf{u})\, \delta\alpha + \frac{\partial}{\partial t}(a \operatorname{div} \mathbf{u})\, \delta t = 0 \tag{13}$$

for elastic bodies and the form

$$\operatorname{grad}_x (a \operatorname{div} \mathbf{v})\, \delta \mathbf{x} + \operatorname{grad}_\alpha (a \operatorname{div} \mathbf{v})\, \delta\alpha + \frac{\partial}{\partial t}(a \operatorname{div} \mathbf{v})\, \delta \mathbf{t} = 0 \tag{14}$$

for fluids.

b) Stability of Compressibility. Compression in the domain d is expressed by the integral

$$\Gamma_d = \int_{d_0} p(\xi, t)\; b(\xi, \alpha, t)\, d\tau_0 , \tag{15}$$

where $p(\xi, t)$ is the pressure and $b(\xi, \alpha, t)$ is a non-homogeneity characteristic concerning the distribution in space and in time of the compression in the neighborhood of the point ξ.

By a similar procedure, writing that the compressibility is stable, hence

$$\delta \Gamma_d = 0 , \tag{16}$$

we obtain the relation

$$\operatorname{grad}_{\mathbf{x}}(b\, p)\delta \mathbf{x} + \operatorname{grad}_\alpha (b\, p)\delta\alpha + \frac{\partial}{\partial t}(b\, p)\delta t = 0, \tag{17}$$

to which must be added the equation of state which connects the pressure p to the density ρ.

c) Relations Expressing the Effects of Viscosity. With the expansion which is expressed by div u and the compression which is expressed by p, hence by the density's equation of state, we have completed the holonomic magnitudes of state. It only remains to find a relation of viscosity corresponding to the flux of matter through the surface of the domain d. In order to obtain for the flux $d\varphi/dn\, d\sigma$ (with $\varphi = u$ for elastic bodies and $\varphi = v$ for fluids) a complete form where account is taken of any possible non-homogeneity of the matter, we introduce a matrix

$$(18) \qquad (A) = I + (\omega_{ij}\,(\xi, \alpha, t))_{i,j=1,2,3}$$

which characterizes the microstructure of the material and I is the unitary matrix.

The expression of the flux becomes then $d\varphi/dn\,(A)d\sigma$. If we insert the temporal part in the form of the normal component W_n of a vector $W(x, t)$, the condition which we have in view may be written

$$(19) \qquad \int_s \left(\frac{d\varphi}{dn}(A)\delta \mathbf{x} + W_n\,\delta t\right) d\sigma = 0 \,.$$

Applying to the above equality Gauss's formula, we obtain

$$(20) \qquad \int_d \left[\sum_h \left(\Delta\varphi_h + \sum_{j,k} \frac{\partial}{\partial x_k}\left(\omega_{jk}\frac{\partial \varphi_j}{\partial x_h}\right)\right)\delta\, x_h + \operatorname{div} W\delta t\right] d\tau = 0$$

$$j,h,k = 1,2,3 \,.$$

Dividing by $\int_d d\tau$ and passing to the limit we obtain the characteristic relation of the viscosity.

$$(\Delta\varphi + \theta)\,\delta \mathbf{x} + \operatorname{div} W.\delta t = 0 \,.$$

This relation becomes

$$(21) \qquad (\Delta \mathbf{u} + \theta)\,\delta \mathbf{x} + \operatorname{div} W.\delta t = 0$$

for elastic bodies, and

$$(\Delta \mathbf{v} + \theta)\delta \mathbf{x} + \operatorname{div} W.\delta t = 0 \tag{22}$$

for fluids.

3. Combining the Foregoing Relations into a Single Expression. By applying the Lagrange method, the foregoing relations concerning elastic bodies as well as fluids may be combined in a single expression with the help of the coefficients

$$\lambda = \lambda(\mathbf{x}, t), \qquad \mu = \mu(\mathbf{x}, t), \qquad \nu = \nu(\mathbf{x}, t),$$

which differ from one material to another, and may or may not vary from one point of the system to another, from one moment to another.

The relation which results by associating (4), (14), (18) has the form

$$D\Omega_\delta^{(i)} + \delta S = \mathbf{A}\delta \mathbf{x} + \mathbf{B}.\delta\alpha - C\delta t = 0, \tag{23}$$

where

$$\mathbf{A} = \rho \frac{d\mathbf{v}}{dt} - \lambda \operatorname{grad}_x (a \operatorname{div} \mathbf{v}) - \mu \operatorname{grad}_x (b\, p) - \nu(\Delta \mathbf{v} + \theta),$$

$$\mathbf{B} = \rho \frac{d\theta}{dt} - \lambda \operatorname{grad}_\alpha (a \operatorname{div} \mathbf{v}) - \mu \operatorname{grad}_\alpha (b\, p),$$

$$C = \rho \frac{dh}{dt} + \lambda \frac{\partial}{\partial t} (a \operatorname{div} \mathbf{v}) + \mu \frac{\partial}{\partial t} (b\, p) + \nu \operatorname{div} W,$$

and concerns fluids.

The relation which results by associating (4),(13),(17) and (10) has the form

$$D\Omega_\delta^{(i)} + \delta S^* = \mathbf{A}^*\delta \mathbf{x} + \mathbf{B}^*\delta\alpha - C^*\delta t = 0,$$

where

$$\begin{aligned} \mathbf{A}^* &= \rho \frac{d^2 \mathbf{u}}{dt^2} - \lambda^* \operatorname{grad}_x (a \operatorname{div} \mathbf{u}) - \mu \operatorname{grad}_x (b\, p) - \nu (\Delta \mathbf{u} + \theta), \\ \mathbf{B}^* &= \rho \frac{d\theta}{dt} - \lambda^* \operatorname{grad}_\alpha (a \operatorname{div} \mathbf{u}) - \mu \operatorname{grad}_\alpha (b\, p), \\ C^* &= \rho \frac{dh}{dt} + \frac{\partial}{\partial t} (a \operatorname{div} \mathbf{u}) + \frac{\partial}{\partial t} (b\, p) + \nu \operatorname{div} \mathbf{W}, \end{aligned} \tag{24}$$

and concerns linear elastic bodies.

§4. The Expression $\Omega_\delta^{(p)}$ for the Potential Field

The potential field is defined by a spatial vector $\mathbf{a}(a_1, a_2, a_3)$ and a scalar potential a_o. The respective elementary potential form is

$$\omega_\delta^{(p)} = \mathbf{a}(\xi, t)\,\delta\xi - a_o(\xi, t)\,\delta t \tag{25}$$

and the form corresponding to the mass included in d is

$$\Omega_\delta^{(p)} = \int_d (\mathbf{a}(\xi, t)\,\delta\xi - a_o(\xi, t)\,\delta t)\, d\mu \tag{26}$$

where the mass element $d\mu$ is of a nature appropriate to that of the field: material mass, electric charge, or any other type of mass.

If we apply on $\Omega_\delta^{(p)}$ the treatment already used in chapter II, we obtain

$$\Omega_\delta^{(p)} = A\,\delta\mathbf{x} + B\,\delta\alpha - C\,\delta t \tag{27}$$

where $A = A(\mathbf{x}, \alpha, t)$, $B = B(\mathbf{x}, \alpha, t)$, $C = C(\mathbf{x}, \alpha, t)$ make up the potential derived from $(\mathbf{a}, a_o)$.

By computing the external differential coefficient of $\Omega_\delta^{(p)}$ we have obtained in chapter II

$$D\Omega_\delta^{(p)} = P.\delta\mathbf{x} + Q\delta\alpha - R\,\delta t, \tag{28}$$

where the components P_j $(j = 1, 2, 3)$ of P, the components $Q_j (j = 1, 2, 3)$ of Q and R have the following expressions

$$\begin{aligned}
P_j &= \frac{dA_j}{dt} - \sum \frac{\partial A_h}{\partial x_j}\frac{dx_h}{dt} - \sum \frac{\partial B_h}{\partial x_j}\frac{d\alpha_h}{dt} + \frac{\partial C}{\partial x_j},\\
Q_j &= \frac{dB_j}{dt} - \sum \frac{\partial A_h}{\partial \alpha_j}\frac{dx_h}{dt} - \sum \frac{\partial B_h}{\partial x_j}\frac{d\alpha_h}{dt} + \frac{\partial C}{\partial \alpha_j},\\
R &= \frac{dC}{dt} + \sum \frac{\partial A_h}{\partial t}\frac{dx_h}{dt} - \sum \frac{\partial B_h}{\partial t}\frac{d\alpha_h}{dt} + \frac{\partial C}{\partial t},
\end{aligned} \tag{29}$$

$$j = 1, 2, 3\,.$$

§ 5. The Equations of Motion

These are obtained as an application of the principles of the invariantive mechanics, from the equality

$$D\Omega_\delta^{(i)} + \delta S^{(i)} = D\Omega_\delta^{(p)} + \delta S^{(e)}, \tag{30}$$

for each $\delta \mathbf{x}$, $\delta\alpha$, δt.

It follows, using (19) and (24), that the equations of motion of a fluid are

$$\begin{aligned} \rho \frac{d\mathbf{v}}{dt} - \lambda \, \mathrm{grad}_x \, (a \, \mathrm{div} \, \mathbf{v}) - \mu \, \mathrm{grad}_x \, (bp) - \nu(\Delta \mathbf{v} + \theta) &= P + P^{(e)}, \\ \rho \frac{d\theta}{dt} - \lambda \, \mathrm{grad}_\alpha \, (a \, \mathrm{div} \, \mathbf{v}) - \mu \, \mathrm{grad} \, (bp) &= Q + Q^{(e)} \end{aligned} \tag{31}$$

and that the equation of energy will take the form

$$\rho \frac{dh}{dt} + \lambda \frac{\partial}{\partial t} (a \, \mathrm{div} \, \mathbf{v}) + \mu \frac{\partial}{\partial t} (bp) + \nu \, \mathrm{div} \, W = R + R^{(e)}. \tag{32}$$

By using the expression of $D\Omega_\delta^{(i)} + \delta S^*$, we obtain for the motion of an elastic body

$$\begin{aligned} \rho \frac{\partial^2 \mathbf{u}}{\partial t^2} - \lambda^* \, \mathrm{grad}_x \, (a \, \mathrm{div} \, \mathbf{u}) - \mu \, \mathrm{grad}_x \, (bp) - \nu \, (\Delta \mathbf{u} + \theta) &= P + P^{(e)}, \\ \rho \frac{d\theta}{dt} - \lambda^* \, \mathrm{grad}_\alpha \, (a \, \mathrm{div} \, \mathbf{u}) - \mu \, \mathrm{grad}_\alpha \, (bp) &= Q + Q^{(e)} \end{aligned} \tag{33}$$

and for the energy

$$\rho \frac{dh}{dt} + \lambda^* \frac{\partial}{\partial t} (a \, \mathrm{div} \, \mathbf{u}) + \mu^* \frac{\partial}{\partial t} (b p) + \nu \, \mathrm{div} \, W = R + R^{(e)}. \tag{34}$$

In the above expressions h represents the proper energy per unit of material mass. Since in the case of ordinary fluids and elastic bodies the velocities involved are very remote from the velocity of light, the energy h reduces to the kinetic energy per unit of mass. We remark also that owing to the specific character of the strain u we may generally put $\partial^2 u/\partial t^2 = d^2 u/dt^2$.

§ 6. The External Constraints

The equations of motion which are derived from (24) are valid if any of the following conditions is fulfilled:

- The external constraints are zero.
- The constraints are included in the expression of the field.

In the first case there is nothing more to add. The second case needs some explanations.

We remark that the integral

$$\int_D (P\ \delta \mathrm{x} + Q\,\delta\alpha - R\,\delta \mathrm{t})\,d\tau$$

which figures both in (26) and in (29) may be considered as a mechanical work in the space of the variables $x_1, x_2, x_3, \alpha_1, \alpha_2, \alpha_3, t$, provided that we ascribe to P, Q, R the significance of a generalized force. It represents the mechanical work effected by the continuous system S under the effect of the field due to the whole mass which has been considered.

But any external constraints acting at the surface of D through the agency of the medium wherein the motion of S takes place, result also in a mechanical work of all the parts of S. This means that the surface integral which represents any mechanical work, done directly by the forces acting on the surface, may be transformed in a volume integral:

$$\int_S (\mathbf{F}\,\delta \mathrm{x} + \mathbf{G}\,\delta\alpha - K\,\delta \mathrm{t})\,d\sigma = \int_D (\mathbf{P}^{(e)}\ \delta \mathrm{x} + \mathbf{Q}^{(e)}\ \delta\alpha - R^{(e)}\delta \mathrm{t})\,d\tau,$$

where $P^{(e)}$, $Q^{(e)}$, $R^{(e)}$ can have only a conventional significance.

The equivalence of the two integrals renders easy the application of the general principle of invariant mechanics.

§7. The Integral Form of the Equation of Motion

The computations developed up to the present concern each part d of the continuous system considered, thus apparently justifying the application (24) of the local principle of invariant mechanics.

It may seem then that the continuous system as a whole intervenes effectively in our theory only by the boundary conditions.

For example, if there were no constraints on the surface of an elastic body, it would appear that the motion of each component d of the body is independent of the motion of the other parts. However, this conclusion is not justified for the following essential reasons:

The internal expansion, compression and viscosity constraints which occur in the foregoing relations are due to the influence of the other parts of the body upon the component part d. If these influences cease, their presence in the equations of motion must cease too, and the respective coefficients λ, or μ or ν must vanish. Therefore these coefficients mark the active presence of all the other parts of the body and their variation shows the greater or smaller effect of the constraints on the component part d.

The above remark, intended in the first place to show that the independence of the different parts of the body is only apparent, is of a greater importance since it compels us to look deeper into the form of the coefficients λ, μ, ν, which are essential characteristics of continuous systems and should be investigated separately by adequate methods concerning the structure of such systems.

The theories concerning stresses in elastic bodies, for instance, may be considered from the viewpoint of invariantive mechanics as intended to supply values of the coefficients λ, μ, ν for the respective bodies.

Furthermore it should be noted that certain plastic systems require for the coefficients λ, μ, ν distributions instead of functions. But this compels us to retain for the equations of motion the integral form where λ, μ, ν may acquire the significance of distributions.

Thus we have to consider the equations of motion in their initial form, which is the integral form, with account being taken of the external constraints

$$\int_D [(A-P-P^e)\,\delta\mathbf{x} + (\mathbf{B}-\mathbf{Q}-\mathbf{Q}^e)\,\delta\alpha - (C-R-R^e)\,\delta t]\,d\tau = 0$$

for fluids and

$$\int_D [(A^* - P - P^e)\delta \mathbf{x} + (B^* - Q - Q^e)\delta\alpha - (C^* - R - R^e)\delta t]d\tau = 0$$

for elastic or, more generally, deformable bodies.

On the Initial Conditions. Since the general aspects of the invariantive mechanics presented here are not concerned with any particular problem we shall not go into a detailed study of the problems set either by initial or boundary conditions.

On Field in General. We have treated in the foregoing of fields which are derived from a potential (a vector plus scalar potential considered generally, in a convenient but improper way a quadrivector).

But in this manner we are far away from the general case treated by classical mechanics.

In the case of a material particle we could envisage the general aspect of the field which is not necessarily derived from a potential, by the following remark which takes us back to what is essential in the fundamental principle, namely the equality

$$\int_C D\omega_\delta^{(i)} = \int_C D\omega_\delta^{(p)}$$

for any closed simple contour C in the space of motion, the displacement along C being designated by δ.

But by definition we have

$$D\omega_\delta^{(p)} = d\omega_\delta^{(p)} - \delta\omega_d^{(p)},$$

where d and δ are symbols of differentiation, which is not necessary for a valid interpretation of the integral $\int D\omega_\delta^{(p)}$. It is sufficient that d as well as δ be the symbols of variation so that one may write only symbolically

$$\delta A_j = A_{j1}\,\delta x^1 + A_{j2}\,\delta x^2 + A_{j3}\,\delta x^3 + A_{jo}\,\delta t$$

without requiring that the second member be an exact differential. The integrals taken along C retain a sense and the final calculation leads to well defined results expressed by the A_{jk} as in the case of § 4, Chap I.

In the theory of continuous systems the passage to the general case of the field no longer derived from a potential is achieved on the same basis by the same procedure, as that applied in the case of a single point.

§8. Influence of Heat

In the classical theory the influence of heat on the motion of a continuous system becomes manifest, besides the presence of an energy vector W, corresponding to a mass force which derives from a potential proportional to the temperature

$$Q = -\chi \operatorname{grad} T .$$

The temperature T constitutes a scalar potential determined at each point of the material mass and at each time t

$$T = T(x_1, x_2, x_3, t).$$

Q is the heat flux per unit of mass.

As has been shown by Fourier, the amount of heat which flows through a surface element per unit time, is proportional to the normal derivative of the temperature and to the surface s; hence its value is

$$-\chi \frac{\partial T}{\partial \varkappa} s,$$

where χ is the coefficient of internal conductivity of the mass. The above expression explains the nature as well as the sign of χ.

CHAPTER VI

THE INVARIANTIVE COSMOLOGY

§1. Introduction

1. The central object of cosmological mechanics is the system S of galaxies and of other material formations as are the quasars or other formations of the same size of the Universe in its totality.

The structure of this system which detaches itself both by the size of the components and by that of the distances which separates them, appreciated by Fred Hoyle as a hundred million light years, that is, about one per cent of the biggest distance accessible to observation till now.

The cosmologists find that at this level one may speak of a homogeneous and isotropic distribution of these bodies, which also may be considered in a first stage as material points in the usual sense of Mechanics.

The study of individual structure mass that makes a galaxy or some other formation of the same level is at the same time the object of the physical sciences and mechanics which come out of the frame of this chapter, as well as physics of dispersed matter in this Universe under the form of radiations of different kinds from electrons to mezons and cosmic rays.

Even the general movement, as for example the rotations of the galaxies, because they are also in an ineraction with other bodies, are not examined here, though elements for such an examination are to be found in §1 of the Vth chapter.

2. The natural way to build up the mechanics of the system S is the direct extension of the invariantive mechanics supported by the fundamental idea of the identical behaviour of the matter at any level it is considered, on the earth or in the sky, in the planet sky first and then in that of the stars, as Galilei and Newton have thought, and then to the level of galaxies which fall under our experiment. The necessary condition is the conformity to the empirical data. But the most verified empirical data that we have in the mechanics of the system S are contained by Hubble's law, according to which the recession speed of any galaxy compared to our own is proportional to distance

$$w = kr \ .$$

The extension of this empirical law to any system of two galaxies has been done on the basis of the cosmological principle of homogeneity and isotropy mentioned above.

Hubble's law is, evidently, a kinematic law only, as Kepler's laws were in the mechanics of our planetary system.

Invariantive mechanics, like that of Newton's, gives the dynamic law of the motion of S; one of the consequences of this law is the conservation of the total impulse of the system from which comes out, without any new hypothesis, a Hubble's type law, justifying also the applicability of the invariantive mechanics.

§2. Invariantive Mechanics of the System S_n

1. We note $P_j(\rho_j, m_j, p_j)$, $j = 1,2,\dots, n$ the n points of the system S_n ; ρ_j and p_j are the respective vectors according to a Euclidean frame of reference, m_j is the dynamic mass $m_j = m_j/\sqrt{1 - v_j^2/c^2}$ and $r_{ij} = \rho_j - \rho_i$ the vector P_iP_j. The motion is represented in the Newtonian space-time.

According to the principles of invariantive mechanics, the function of state E, which is the energy of the system, is a Euclidean invariant of it. So it is a function of the invariant quantities which defines the state of the system, except a Euclidean motion of the frame of reference.

These quantities are

$$\alpha_j = \frac{1}{2}\pi_j^2 \; ; \quad \alpha_{jk} = \pi_j \pi_k , \quad (j<k) \; ; \quad \beta_{jk} = \pi_j r_{jk} \, (j \neq k),$$

$$\gamma_{jk} = \pi_k r_{jk} \; ; \quad \delta_{jk} = \frac{1}{2} r_{jk}^2 \, (j \neq k), \tag{1}$$

$$\epsilon_{jhk} = r_{jk} \, r_{kh} \, (j \neq h \neq k) , \quad j,h,k = 1,2,\dots,n \; .$$

We shall have

$$E = E(\alpha_j, \alpha_{jk}, \beta_{jk}, \gamma_{jk}, \delta_{jk}, \epsilon_{jhk}). \tag{2}$$

2. The principle that may be applied shows that

$$J^{(i)} = \int_L \omega_\delta^{(i)} \, , \text{where } \omega_\delta^{(i)} = \sum \pi_j \delta\rho_j - E\delta t \tag{3}$$

is an invariant of the movement we represented by the operator d. This gives

$$DJ^{(i)} = \iint_{L \times \ell} d\omega_\delta^{(i)} - \delta\omega_d^{(i)} = 0 \tag{4}$$

But

$$d\omega_\delta^{(i)} - \delta\omega_d^{(i)} = \sum \delta\pi_s d\rho_s + \sum d\pi_s \, \delta\rho_s - dE\,\delta t + dt\,\delta E, \tag{5}$$

where

$$\delta E = \sum E_{\alpha_s} \pi_s \delta\pi_s + \sum E_{\alpha_{sk}} (\pi_s \delta\pi_k + \pi_k \delta\pi_s) +$$

$$+\sum E_{\beta_{sk}}\,\pi_s\,\delta r_{sk} + \sum E_{\beta_{sk}}\,r_{sk}\,\delta\pi_s + \sum E_{\gamma_{sk}}\,\pi_k\,\delta r_{sk} +$$

$$+\sum E_{\gamma_{sk}}\,r_{sk}\,\delta\pi_k + \sum E_{\delta_{sk}}\,r_{sk}\,\delta r_{sk} + \sum E_{\epsilon_{jhk}}(r_{jk}\,\delta r_{kh} + r_{kh}\,\delta r_{jk}) =$$

$$= \sum (E_{\alpha_s}\,\pi_s + \sum E_{\alpha_{sh}}\,\pi_h + \sum E_{\beta_{sh}}\,r_{sh} + \sum E_{\gamma_{hs}}\,r_{hs})\,\delta\pi_s +$$

$$(6)\qquad + \sum E_{\beta_{jk}}\,\pi_j\,(\delta\rho_k - \delta\rho_j) + \sum E_{\gamma_{jk}}\,\pi_k\,(\delta\rho_k - \delta\rho_j) +$$

$$+ \sum E_{\delta_{jk}}\,r_{jk}\,(\delta\rho_k - \delta\rho_j) + \sum E_{\epsilon_{hkj}}\,r_{hj}(\delta\rho_k - \delta\rho_j) +$$

$$+ \sum E_{\epsilon_{jhk}}\,r_{kh}\,(\delta\rho_k - \delta\rho_j)$$

or also

$$(7)\qquad \delta E = \sum \Phi_s \delta\pi_s + \sum L_s\,\delta\rho_s\,,$$

with

$$L_s = \Psi_s\,\pi_s + \sum \Psi_{sk}\,\pi_k + \sum \chi_{ks}\,r_{ks} + \sum \chi_{hks}\,r_{hk}\,,$$

$$\Phi_s = E_{\alpha_s}\,\pi_s + \sum E_{\alpha_{sh}}\,\pi_h + \sum F_{sh}\,r_{sh}\,.$$

To write that $DJ^{(i)} = 0$ for any δ comes to ask that (5), where we take into account (6), be cancelled, after dividing by dt; that is, that

$$\sum\left(\frac{d\pi_s}{dt} + \Psi_s\pi_s + \sum \Psi_{ks}\,\pi_k + \sum \chi_{ks}\,r_{ks} + \sum \chi_{hks}\,r_{hk}\right)\delta\rho_s -$$

$$(8)\qquad - \sum_{s=1}^{n}\left(\frac{d\rho_s}{dt} - \Phi_s\right)\delta\pi_s - \frac{dE}{dt}\;\delta t = 0.$$

From here it comes out the following system of equations

$$(8)\qquad \frac{d\rho_s}{dt} = \Phi_s\,,$$

$$\frac{d\pi_s}{dt} + L_s = 0 \qquad (s = 1, 2, \ldots, n) \tag{8}$$

and

$$E = E_o . \tag{9}$$

The system (7) defines the impulse π_s according to velocities and positions.

The first group gives the equations of motion. The last equation, (9), corresponds to the energy conservation.

According to the formation of the terms of equations (8), it comes out immediately that

$$\sum_{j=1}^{n} \frac{d\pi_s}{dt} = 0 .$$

So

$$\sum_{s=1}^{n} \pi_s = C^t , \tag{10}$$

equation that represents the conservation of the impulse

In order to simplify the writing we may put

$$v_s = \frac{d\rho_s}{dt}$$

The equations (7) which may also be written

$$v_s - \sum F_{sh}\, r_{sh} = E_{\alpha_s}\, \pi_s + \sum E_{\alpha_{sh}}\, \pi_h$$

have the coefficients in π_h symmetrical. Supposing we have

$$\begin{vmatrix} E_{\alpha_{11}} & E_{\alpha_{12}} & \ldots & E_{\alpha_{1n}} \\ E_{\alpha_{21}} & E_{\alpha_{22}} & \ldots & E_{\alpha_{2n}} \\ \ldots & \ldots & \ldots & \ldots \\ E_{\alpha_{n1}} & E_{\alpha_{n2}} & \ldots & E_{\alpha_{nn}} \end{vmatrix} \neq 0 ,$$

the equations (11) have a solution like

$$\pi_j = m_j (v_j - \sum F_{jh} r_{jh}) + \sum \mu_{jl} (v_l - \sum F_{lh} r_{lh}) \tag{11}$$

where $\mu_{jl} = \mu_{lj}$, or also

$$\pi_j = m_j v_j + \sum \mu_{jl} v_l - m_j \sum F_{jh} r_{jh} - \sum \mu_{jl} r_{jh} F_{lh} . \tag{12}$$

If we compare with the expressions got for impulses in the case n = 2, we must consider the coefficients noted with m_j as representing the respective masses

$$m_j = m_j^0 / \sqrt{1 - v_j^2 / c^2} ,$$

and μ_{jl} as the gravitational interaction mass between P_j and P_l,

$$\mu_{j\ell} = -\tfrac{1}{2} f m_j^0 m_\ell^0 \left(1 + \frac{k}{c^2 r_{j\ell}}\right) / c^2 r_{j\ell} \sqrt{1 - v_j . v_\ell / c^2} ; \tag{13}$$

the symmetric coefficients F_{jh} will be characteristics of the dilatational interaction.

3. The Expression of Energy. We find out from the above results that the inertial motion of n bodies does not cause the appearance of other interactions except those which we found in the case n = 2, that is, the gravitational and dilatational one. So we can write as in the case n = 2,

$$E = c^2 (\sum m_j + \sum \mu_{jk} + \sum \nu_{jk}) . \tag{14}$$

In order to find the expressions of interaction masses ν_{jk}, we proceed as in the case n = 2 and find

$$\nu_{jk} = -\tfrac{1}{2} g r_{jk}^2 / c^4 \sqrt{1 - u_{jk}} , \tag{15}$$

with $u_{jk} = [m_j^0 v_j^2 \cos^2 (v_j, r_{jk}) + m_k^0 v_k^2 \cos^2 (v_j, r_{jk})] / 2(m_j^0 + m_k^0) c^2$.

4. The expressions of the impulses are thus the following:

$$\pi_j = m_j \mathbf{v}_j - \sum \mu_{jk} \mathbf{v}_k + \sum h_{jk} \nu_{jk} \mathbf{r}_{jk} , \tag{16}$$

where

$$h_{jk} = m_j^0 \mathbf{v}_j . \mathbf{r}_{jk} / (m_j^0 + m_k^0) r_{jk} . \tag{17}$$

If we take into account all these, the theorem of impulse conservation gives us the equality

$$m_1 \mathbf{v}_1 + m_2 \mathbf{v}_2 + \mu_{12} \mathbf{v}_1 + \mu_{21} \mathbf{v}_2 + h_{12} \nu_{12} \mathbf{r}_{12} + h_{21} \nu_{21} \mathbf{r}_{21} =$$

$$C - \sum_{j>2} m_j \mathbf{v}_j + \sum_{i+j>3} \mu_{ij} \mathbf{v}_j - \sum_{i+j>3} h_{ij} \nu_{ij} \mathbf{r}_{ij} = \Gamma ,$$

where the vector $\mathbf{r}_{12}$ and $\mathbf{r}_{21}$ does not explicitly appear in the second part. We put

$$m_1 \xi_1 + m_2 \xi_2 + \mu(\xi_1 + \xi_2) + (h_{21} + h_{12}) \nu \frac{\mathbf{r}^2}{r^2} = \Gamma \cos(\Gamma, \mathbf{r}) , \tag{18}$$

where

$$\xi_1 = \nu_1 \cos(\mathbf{v}_1, \mathbf{r}), \quad \xi_2 = \nu_2 \cos(\mathbf{v}_2, \mathbf{r}) .$$

If we use the expressions of h_{12} and h_{21} and of μ and ν and if we divide by Γ cos (Γ,$\mathbf{r}$), we obtain

$$\frac{1}{c^2 r} M + \frac{r^2}{c^4} N = 1 , \tag{19}$$

where M and N are expressions which do not contain any more explicitly the vector $\mathbf{r}$ or its value r. They contain all the other distance-vectors, the velocities of the other bodies and angles.

If we derive on both sides of (19) according to time and then we divide by $2r/c^2$, we get

$$(N - \frac{c^2}{2r^3} M) \frac{\mathbf{r} \cdot \dot{\mathbf{r}}}{r} = - \frac{1}{2} (\dot{N} + \frac{c^2}{r^3} \dot{M}) r . \tag{20}$$

But

$$\left| \frac{\mathbf{r}.\dot{\mathbf{r}}}{r} \right| = \frac{| \mathbf{r}(\mathbf{v}_2 - \mathbf{v}_1) |}{r} = | \xi_2 - \xi_1 |$$

is the recession velocity of the bodies P_1 and P_2 which are, in fact, two arbitrary bodies of the system S_n. Formula (20) gives then the theoretical law of dilatation for the distance of the bodies P_1 , P_2 in the system S_n :

$$(21) \qquad W = \frac{1}{2} \frac{|\dot{N} + \frac{c^2}{r^3} \dot{M}|}{|N - \frac{c^2}{2r^3} M|} r .$$

5. The existence of the limit c for the velocities of the material bodies is included in the construction of invariantive mechanics from its first step.

A condition for the validity of the law (21) is necessarily given by the inequality

$$(22) \qquad \frac{1}{2} \frac{|\dot{N} + \frac{c^2}{r^3} \dot{M}|}{|N - \frac{c^2}{2r^3} M|} r < c$$

for any pair of galaxies, thus any r in the system.

6. A first examination in the light of the above considerations and especially with regard to Hubble's empirical law, shows us that we may not have either $N = 0$, which would give

$$W = |\dot{N} + \frac{c^2}{r^3} \dot{M}| \frac{r^4}{|Mc^2|},$$

nor $\dot{N} = 0$, which would give

$$W = \frac{1}{2} \frac{|c^2 \dot{M}|}{|N - \frac{c^2}{2r^3} M|} \frac{1}{r^2},$$

contradictory to the dilatation law.

So we may give the law (21) under the form

$$(23) \qquad W = \frac{1}{2} \frac{|\frac{\dot{N}}{N} + \frac{c^2}{r^3} \frac{\dot{M}}{N}|}{|1 - \frac{c^2}{2r^3} \frac{M}{N}|} r ,$$

which is simpler for the following considerations.

§3. Finiteness or Infiniteness of the Universe

In any case, that the universe has a finite or an infinite number of galaxies, we must extend relation (23) to the whole universe. It is only so that it can get a physical sense and not only a formal one. The bodies that form it exert an ever-powerful ineraction as r is greater, so the formula (9) is exact only if it may imply the presence of all galaxies.

1°. If the universe is finite, we take n equal to the total number of galaxies. In this case M and N have some values which represent the whole system from which we have taken the two bodies P_1 and P_2. We have to suppose a certain stability of the system, and by this that $\dot{M}/N$, $\dot{N}/N$, M/N are limited superior and inferior during a certain period in order to draw the conclusion that, as soon as r exceeds a certain value, the law (9) reduces to the following

$$w = \frac{1}{2}\left|\frac{\dot{N}}{N}\right| r, \qquad (24)$$

where $\dot{N}/N$ does not explicitly depend on r.

The exigence represented by formula (8) becomes in this case

$$r < 2c/\left|\frac{\dot{N}}{N}\right|,$$

inequality which would show the upper limit of the distances among galaxies of the universe because N, so N too, concern the system of all galaxies, have thus a value, which in fact is representing Hubble's constant.

The above mentioned considerations must get two principal corrections included in the following remarks.

Remark 1. N/N is calculated for the whole system of galaxies from which there are taken the two which are distant with r; thus it depends, if not explicitly but implicitly, on r. Being given the very great number of dependent galaxies, this is weak and cannot be sensible but for r which is very big. This shows that the relation $w < c$ may be observed, and the fact that Hubble's law must get a correction compatible both with experience and theory.

Remark 2. The coefficient $\dot{N}/N$ is weakly variable with r for the reason already shown. But it is variable with time by its structure. As M and N which appear in (19) are characteristical sizes of the Universe – indirectly and weakly dependent on r – their derivatives $\dot{M}$ and $\dot{N}$ are two representative velocities of the dynamics of this universe. The experience that confirm Hubble's law especially suggests to us the hypothesis that the variation in time of these derivatives and mostly of the quotients $\dot{M}/M$, $\dot{N}/N$ and of the ratio M/N is little. The period during which our cosmic experiments have been done and particularly the relative measures to Hubble effect is too short to justify by itself the assignment of stability which we have done above. The more that under our telescopes eyes were a lot of cosmic catastrophes of huge dimensions which are of interest for the above mentioned theories, or there appeared bodies in rapid evolution and not yet explained which may disturb any theory. The answer does not seem difficult. First, the time of observations is not only that of our observations here on the Earth, but it is that of superpositions of the duration of communications which we record and which range to billions of light-years. The catastrophies which happened since the remotest galaxies have sent the information about them, by light waves, are local or not so significant according to the number of information we record from all bodies of the universe. So, we must consider, all cosmologists together, that the duration of our experiments is also to the level of the object under observation.

Dilatation or Oscillation?

We deduced the law (21) from formula (20) restricting its applicability only to the equality of the absolute values of the two terms. If we do not do this restriction suggested by the empirical law of dilatation and consider the proper recession with its sign, formula (20) gives us

$$W = \frac{1}{2}\frac{\dot{N}}{N}\,\frac{1+\dfrac{c^2}{r^3}\dfrac{\dot{M}}{\dot{N}}}{1-\dfrac{c^2}{2r^3}\dfrac{M}{N}}\,r \tag{25}$$

or, under the conditions mentioned above

$$W \sim \frac{1}{2}\frac{\dot{N}}{N}\,r$$

which shows that W may be positive or negative, having the sign of the quotient $\dot{N}/N$. Given that variation of this quotient is cosmically slow, the change of its sign would be slow, if this change may take place. In that case we could have during the cosmic time a sequence of motions of dilatations and contractions of the system of galaxies, that constitutes the isles of concentration of the material in the radiation's ocean about which the previous theory has not said anything till now.

2°. The Hypothesis of an Infinite Number of Galaxies. Dilatation problem. It is not easy to accept the idea that the universe of galaxies is finite; although the invariant mechanics supports it, as we shall see.

With regards to the possible validity of formulae (21) and (23), in the hypothesis of an infinite universe, are to be done the following considerations.

First we observe that in Hubble's empirical law there are included the dilatational effects of all bodies of the finite or infinite universe. This proves that in formula (22) in which M and N depend on the number n of the galaxies, we make n to tend to infinite so that finally it may cover all of them, the expression from the second member must converge to a limit which will have the form

$$W = \frac{1}{2} \frac{\left|H + \frac{c^2}{r^3} K\right|}{\left|1 - \frac{c^2}{2r^3} L\right|} r, \tag{26}$$

where H, K, L, are limited at least in the present epoch of the universe life, in the meaning we cleared above.

For r sufficiently big, the law becomes the approximate form

$$W = \frac{1}{2} |H| r, \tag{27}$$

H not being rigorously constant, both as regards time and dependence on r, and being obliged to verify the condition

$$|H| < 2cr^{-1}.$$

Energy Expression. The energy of a galaxy, for example of the galaxy P_1, from the system S_n has the following expression

$$E_1 = m_1 c^2 - f \sum_{j=2}^{n} \frac{m_1^0 m_j^0}{r_{1j}\sqrt{1 - \mathbf{v}_1 \cdot \mathbf{v}_j / c^2}} + \frac{g}{c^2} \sum_{j \geqslant 2} \ell_{1j} r_{1j}^2 / \sqrt{1 - u_{1j}}, \tag{28}$$

where f and g are universal constants, f Newton's constant and g a not yet determined constant.

This expression is only conventional, since we do not consider the system S_n as being made up from all galaxies of the universe, for the interaction of each of them with P_1 being an energetic contribution.

If n is finite the expression of E_1 keeps a finite value, but if n would be infinite, the value of the dynamic energy of the galaxy P_1 would be infinite. This fact we cannot admit. We are then in the position to renounce the hypothesis of an infinite number of galaxies, or to modify the theory which gave a concordance as much complete with the experiments.

Thus we must consider as finite the number of the galaxies and, together with them, the number of all the formations to which they are comparable.

§4. Newtonian Interpretation. The Two Kinds of Inertial Forces

The energy expression given by (28) confirms the simple interpretation of the dynamic structure of the universe, in approximative but suggestive Newtonian language which we have adopted in the classical case.

If, at the same time, we leave the factors $1/\sqrt{1 - \mathbf{v}_1 \cdot \mathbf{v}_j/c^2}$ and $1/\sqrt{1 - u_{1j}}$ aside, we notice that the additive terms in the expression (14) are of two kinds: gravitational potentials $f\, m_1^0\, m_j^0/r_{1j}\,(1 + K_{1j}/c^2\, r_{1j})$ and elastic potentials $g/c^2\, \ell_{1j}\, r_{1j}^2$. The former ones correspond to the gravitational forces, the latter one to an elastic interaction proportional to the distance. These two are but the only potentials compatible with matter.

On Hypotheses Used in the Former Extension of Cosmology. From the beginning, together with the problem of the two bodies, the invariantive mechanics has been obliged to be a science of the universe, as it included, as a consequence of the principles, the terms corresponding to the elastic potential whose sizes are of cosmological dimensions.

A proper cosmological extension, did not consist in adding some new principles or mechanical hypotheses, but only in the act of selecting bodies to which motion is referring to. We have considered, especially galaxies or similar formations, leaving aside the rest of matter which is dissociated in radiations or other forms less known and spread over the universe.

That the system of galaxies is isolated as it is form a mechanical system under the principles of invariantive mechanics, clearly comes from the preceding developments. The homogeneity and isotropy – that is what is called cosmological principle – imposed by the observations and by Hubble's empirical law, serves a posteriori, to the interpretation of the theoretical results that have been obtained.

§ 5. The Relativistic Cosmology

Another quantitative way of supporting the studies of the cosmologic problems was the geometric relativistic one, initiated by Einstein and De Sitter.

In order to present this method in a clearer way, which excessively simplifies the image of the universe, we adopt Einstein's point of view to which he stopped after many hesitations, and expressed in the chapter "On Cosmologic Structure of the Space", of the volume published by Hermann in 1933.

"For simplicity reasons we shall leave aside", says A. Einstein, "the fact that matter is concentrated in the stars and systems of stars, apparently separated by bare spaces, and let's consider it as if being distributed continuously on big cosmic spaces".

It is also said "that the density of ponderable matter energy exceeds in a great extent that of radiation so that the last one may be neglected. Of course, this assumption is not exact in the whole, but the approximation introduced in this way does not change anything essential in the considerations and results that follows".

The representative geometric universe of the continuous matter distribution with a uniform spatial density at any moment cannot be Euclidean, says Einstein, because Euclidean means zero curvature, and the null curvature, according to relativity principles, means null gravitational mass.

Here is the argument which substantiates that which Einstein considers as metrics of the universe.

"The simplest possible structure of the space, after the Euclidean one, seems to be the static one (that is the coefficients of the respective quadratic form must be independent of time) and having a uniform curvature in its spatial sections", and "a three dimensional space with a constant positive curvature (in particular spheric space) is, as it is known, characterized by the linear element, whose form is

$$d\sigma^2 = \frac{dx_1^2 + dx_2^2 + dx_3^2}{(1 + D^{-2}\, r^2)^2} \qquad \text{with } r^2 = x_1^2 + x_2^2 + x_3^2 ,$$

D being the diameter of the sphere supposed to be constant.

"A static and spheric universe from the spatial point of view is, consequently, described by the linear element.

$$ds^2 = \frac{dx_1^2 + dx_2^2 + dx_3^2}{(1 + D^{-2}\, r^2)^2} - c^2\, dt^2 . \tag{1}$$

This, "Consequently" seems to us unjustified, as, for reasons that we shall only mention, A. Einstein replaces the element (1) by the following one:

$$ds^2 = \left(\frac{D}{D_0}\right)^2 \frac{dx_1^2 + dx_2^2 + dx_3^2}{(1 + D_0^{-2} r^2)^2} - c^2 dt^2, \tag{2}$$

where the galactic diameter $D = D(t)$ is a function of t with $D(o) = D_0$; $A = D/D_0$ is "the expansion factor".

The obligation of renouncing to the formula (1) came from the incompatibility of this linear element with the general relativity equations and the constant density hypothesis.

In the third state of the construction of its theory A. Einstein ascertains that "in the actual state of knowledge the fact of the density of matter different from zero must not be theoretically bound with a spatial curvature but with a spatial extension. "This finding is bound to the fact that we may renounce to the spatial curvature and consider for the linear form the expression

$$ds^2 = A^2 (dx_1^2 + dx_2^2 + dx_3^2) - c^2 dt^2$$

where the expansion factor A is only a time function.

If we apply now to this linear form the equations of general relativity, we get an equation which shows that the expansion factor A is a function of the time only and that

$$\frac{\dot{A}}{A} = \frac{\dot{D}}{D}$$

is a constant, which if we make equal to Hubble's constant, it results exactly in the dilatation law of the universe:

$$\dot{D} = hD.$$

It is true that here the law is referring only to the universe in its whole and not only to the distance between any two galaxies, as says Hubble's law in its general form.

In order to find again the general law we must go back to the form (2) of the linear element, as other authors did, and find that one which corresponds to the result we want to get. In fact, the same as we did, following the above mentioned way.

The whole geometric experience showed us that the expression of a ds^2, generally, gives a local information and that there are very few of the Riemannian varieties for which a given local ds^2 may give an information for the whole variety, as we want for the linear form which is to represent the universe with its whole past, even with its birth history, that is with a moment when it is in a singular state.

Without allowing ourselves a judgement on cosmologic theories generally gathered under the name "stationary state" with the changes brought ceaselessly under the observations empire and the progresses in physics and chemistry, does not seem to exist in a kind of incompatibility with the results brought by invariantive mechanics which deals only with global movements of matter's agglomerations which make up the galaxies and for the moment leaves aside both the physical and mechanical structure of these systems considered in its microstructure and its relations with the ocean of radiations where they are forming and moving.

COMMENTARY 1

SPACES AND AXIOMS OF MOTION

1. Introduction

The discussion which follows purports to examine from a more precise axiomatic point of view the notions that have been employed and particularly the spaces and the principles.

Only after being in possession of formalized elements can we state precisely the principles of the Invariantive Mechanics of the motion of a material particle either in a form which has become classic, although it involves a variable mass, or in a more general field of a tensorial character.

This is in general the role of axioms: they help in the first place to express precisely the elements of a theory already verified in order to gain a solid support for natural generalizations.

We have employed in the preceding pages different spaces without stating their respective properties.

Now we must re-examine the part played by each of these spaces and state the relations between them.

We shall see on this occasion that the usefulness of the Einstein-Minkowski space does not precede but follows the theory of the inertial motion of a material particle, that this space is necessary only when we pass to the motion in a field, since the structure of the Einstein-Minkowski space serves to obtain an adequate geometrical structure of the field.

Or rather, it serves to obtain a geometrical interpretation of the analytical structure which is also the outcome of physical theories.

2. The Spaces

Let us first specify the different spaces considered.

a) The space S which is the three-dimensional Euclidean space of Newton. It is the space wherein bodies are located and which is used in the sciences of nature and technics.

The group of Euclidean transformations of this space consists of translations and rotations which preserve distances and angles. There is no Euclidean scalar invariant corresponding to a unique point of this space. There exists a unique invariant corresponding to a system of two points, which is the distance r.

b) The space of points T; this is a one-dimensional Euclidean space identical to the time-line of the classical mechanics and astronomy.

The group of Euclidean transformations of the space T consists of translations.

c)The three-dimensional vectorial space $S'(\equiv E_3)$isomorphic to the space of the pairs (P_0, P), where P_0 is a fixed point and P is a generator point of S.

The Euclidean invariants associated with a vector **u** of S' are $\mathbf{u}.\mathbf{u} = u^2$ as well as any scalar function of $\mathbf{u}^2$ which for convenience in calculations shall be denoted by $f(1/2\, u^2)$.

The Euclidean scalar invariants of a system of two vectors **u** and **v** of S' are u^2, $\mathbf{u}.\mathbf{v}$, v^2 and any scalar function of the preceding invariants $f(1/2\, u^2, \mathbf{u}.\mathbf{v}, 1/2\, v^2)$.

d) The space $T'(\equiv E_1)$ isomorphic with the space of the differences $t - t_0$, where t_0 is a fixed value.

e) The four-dimensional punctual space R defined as the Cartesian product of S and T

$$R = X \times T.$$

f) The vectorial space $R' = S' \times T'$.

g) The seven-dimensional phase-time space U, defined by the Cartesian product

$$U = S \times E_3 \times T,$$

where E_3 is the Euclidean three-dimensional space.

h) The space $U' = S' \times E_3 \times T'$.

3. The Invariants of the Spaces R and U

The only transformations which leave invariant the structure of each of the two spaces are those which refer separately to S(S'), E_3 and T(T'). The first two are identical to the Euclidean transformations defined on S, provided we identify E_3 to the space of the pairs of points P_0, P of S. The Euclidean transformations defined on S have no effect on E_3 or T.

It follows that the transformations to be considered on R consist of the set of Euclidean transformations on S associated with the transformations on T. The latter have no invariants other than the absolute constants.

This shows that the only time independent scalar invariants of the transformations to be considered on R and U are those of the space S.

4. The Differential Operator δ and the Corresponding Differential Forms

The differential operator δ is defined as follows:

1°. Let $\mathbf{P}(x_1, x_2, ..., x_n)$ be a point of the numerical space R_n and let $\delta\mathbf{P}(\delta x_1, \delta x_2, ..., \delta x_n)$ be the generic vector of the vectorial space R_n identical with R_n as vectorial space. To each specified δ and to each $\mathbf{P}$ there corresponds a vector $\delta\mathbf{P}$ of R_n and only one, and, conversely, to each vector of R_n there corresponds a $\delta\mathbf{P}$ and only one.

2°. The effect of δ upon a scalar $F(\mathbf{P})$ of R_n is expressed by the scalar product of the vector grad $F(\mathbf{P})$ of R and the vector $\delta\mathbf{P}$ which corresponds, as shown in the previous identification, to a generic vector of R_n :

$$\delta F(\mathbf{P}) = \text{grad } F . \delta\mathbf{P},$$

provided, of course, that grad F exists.

In general, we postulate that the rules for the application of δ to sums or products, to vectors or vectorial products associated with the points of R_n are those of ordinary differentiation, taking into account the characteristics of $\delta\mathbf{P}$ as indicated at point 1°.

3°. A specified δ , such as d for instance, will be defined in the sequel by a vectorial equality

$$d\mathbf{P} = \mathbf{V} . ds,$$

where $\mathbf{V}$ is a well defined vector associated with the point P of R_n, and ds is an arbitrary numerical value.

This definition shows that the vector $\mathbf{V}$ is equal almost overall along a curve $\mathbf{P} = \mathbf{P}(s)$ to the vectorial differential coefficients $d\mathbf{P}/ds$ calculated at the point $\mathbf{P}$.

The parameter s may be identified, when it is necessary, to t.

We must consider also expressions of the form

$$A_\delta = \sum_{j=1}^{n} A_j \; (\mathbf{P}) \, \delta x_j \; ,$$

which are, or are not, scalar products according to whether the system $A_j(P)$ is, or is not, vectorial.

5. Vectorial Spaces and Generated Vector Fields

The general differential operator δ establishes a correspondence between a point $M(\mathbf{x},t)$ of the space R, and the four-dimensional space of the vectors $\delta M(\delta\mathbf{x},\delta t)$ with the components $\delta\mathbf{x},\delta t$; the latter may be identified to the space R.

The same operator δ establishes a correspondence between a point $N(\mathbf{r},\mathbf{u},t)$ of the space U, and the seven-dimensional space which may be identified to U′ and will be denoted $U_{r,u,t}$.

The operator δ generates further two fibred spaces* $\tilde{R}$ and $\tilde{U}$ defined as follows

$$\tilde{R} = \{R_{r,t} \mid (\mathbf{r},t) \in R\}$$

$$\tilde{U} = \{U_{r,u,t} \mid (\mathbf{r},\mathbf{u},t) \in U\} .$$

6. The Differential Forms Defined by the Generic Operator δ.

Let $\mathbf{u}$ be a vector of S′ and u_0 a vector of T′. Following obvious identifications we consider in $R_{r,t}$ the systems $\mathbf{u}$, u_0 and δM whose components are δr, δt. The first is a fixed vector and the second a generic vector. The expression

$$\omega_\delta = \mathbf{u}.\delta\mathbf{r} + u_0\,\delta t$$

is a sum of the scalar products $\mathbf{u}.\,\delta r$ and $u_0\,\delta t$ of the respective vectorial spaces S′ and T′, hence it is an invariant under the most general Euclidean transformation of $U_{r,t}$ in which ω_δ is defined. Indeed, this transformation operates separately in S′ and T′. It leaves invariant $\mathbf{u}.\delta\mathbf{r}$ if $\mathbf{u}$ depends only on $t-t_0$ as well as $u_0\,\delta t$ if u_0 is a scalar invariant under the transformations of S′ and of the time vectors δt.

These conditions show that the most general invariant form of the Euclidean transformations of R are obtained by taking

$$u_0 = f\left(\frac{1}{2}\,u^2\right).$$

---

* The concept of fibred space is explained in Commentary 2.

Similarly, it may be seen that θ_δ, associated with U and invariant under the Euclidean transformations of R′, will be of the form

$$\theta_\delta = \mathbf{u}.\delta\mathbf{r} + \mathbf{v}.\delta\pi + w\delta t,$$

where $\delta r \in E_3$, $u \in E_3$, $v \in E_3$ and w is a constant with respect to t and a function of the invariants $\mathbf{u}^2$, u.v, $\mathbf{v}^2$, hence

$$w = f(\frac{1}{2}\, u^2 , \mathbf{u}.\mathbf{v}, \mathbf{v}^2).$$

θ_δ is invariant under the Euclidean transformations of E_3 being the sum of three invariant expressions.

Remark 1. If $\theta_\delta = 0$ for a generic δ we obtain necessarily $u = 0$, $v = 0$, $w = 0$.

Remark 2. For the specified operator d the external derivative of ω_δ is an invariant under the Euclidean transformations of S′ which are independent of time.

Indeed we have

$$D\omega_\delta = \mathbf{du}\,\delta\mathbf{r} - (\mathbf{dr} + \mathrm{grad}\; f dt)\delta\mathbf{u} + df\delta t \;.$$

But it is obvious that **du** belongs to S′ as does **u** and **dr**, grad f, df and f itself do not depend on t and are invariant under the Euclidean transformations of S′.

7. Inertial Motion of the Material Particle. Definition 1.

A moving material particle is an element of a vector field belonging to the fibred space Σ consisting of the association of each point M(r,t) with a vector **I** (impulse or momentum) of S′. Considering the equivalence between Σ and the space U, we may defined the material particle as a point (r,**I**,t) of the phase-time space U. The motion of the material particle will be defined as a mechanical category after we shall have stated the postulates of Mechanics.

We may, however, define the motion of a material particle only as a geometrical category in order to achieve a first important deliminitation of this object. For this reason the definitions which follow, will be called delimitations.

Delimitation 1. The motion of the material particle (r,**I**,t) is a continuous mapping of the half straight line $t \in T$, $t \geqslant t_0$, into the space U.

Therefore it has a support $\Gamma(r = r(t),\ \mathbf{I} = \mathbf{I}(t),\ t(\geqslant t_0))$ with a continuous law of displacement on the support, starting from $t = t_0$.

Delimitation 2. Motion in the ordinary sense is the restriction $\mathbf{r} = \mathbf{r}(t)$, $t \geqslant t_o$ on the space R of the motion considered above.

The delimitation 2 is necessary since the first delimitation which requires that $\mathbf{I}(t)$ be continuous imposes on $\mathbf{r}(t)$, on which $\mathbf{I}(t)$ is dependent, a condition much stronger than simple continuity.

Remarks on the preceding delimitations. So far we have given for motion simple delimitations; for this reason we may call it geometrical motion.

But the motions of mechanics are natural motions, facts of experience, which are much more particular than the geometrical motions among which they constitute a particular class.

For this reason the above delimitations are useful as they offer a field of elements wherein the principles of mechanics must operate a choice of their objects.

Principles and Postulates

1. Generalities. We know that the most general invariant differential form associated with the fibred space Σ and with $\mathbf{I}$ is expressed by

$$\omega_\delta^{(i)} = \mathbf{I}.\delta \mathbf{r} - f(\tfrac{1}{2}\,\mathbf{I}^2)\,\delta t \;.$$

We recall that $\omega_\delta^{(i)}$ cannot be considered yet as a scalar product in R or R′, since these spaces are not endowed with a metrics; $\omega_\delta^{(i)}$ is only the sum of two scalar products of the spaces S′ and T′.

We also know that the geometrical derivative of $\omega_\delta^{(i)}$ is itself an invariant form in $U_{r,I,t}$ under Euclidean transformations of E_3, and that

$$D\omega_\delta^{(i)} = d\mathbf{I}.d\mathbf{r} - (d\mathbf{r} - \mathrm{grad}\, f.dt)\,\delta\mathbf{I} - df.\delta t \;.$$

We can now determine the motion and hence the operator d with the help of one principle and three postulates of which the second is of a special character.

Principle. Motion cancels the geometrical derivative of $\omega_\delta^{(i)}$ for any operator δ.

Postulate 1. The ratio f/m (where $m = 1/\dot{f}$,) is a universal constant.

We are led to put $f = \epsilon m \omega^2$, $m = m_o/\sqrt{1-\epsilon\, v^2/\omega^2}$.

Postulate 2. We take $\epsilon = +1$ and may call this the Einstein–Minkowski postulate.

Postulate 3, or the Einstein postulate:, $\omega = c$ = velocity of light in vacuum.

The mechanics of a material particle in inertial motion is thus defined.

The function of state $f = mc^2$ represents the energy of the mass in motion.

2. Universe and metrics. One of the important results of the preceding axiomatic theory is that the metrization of the space R becomes possible.

Indeed, we may write

$$\omega_\delta^{(i)} = m(\mathbf{r}.d\mathbf{r} - c^2 dt^2) = - m_0 c\sqrt{c^2 dt^2 - dr^2} \ .$$

The second member emphasizes the pseudoeuclidean metrics

$$d\sigma^2 = c^2 dt^2 - dr^2$$

of the space R. Endowed thus with a metrics this space will be called the Einstein-Minkowski space of universe and denoted in the sequel by R_{EM}.

The axiomatic construction of the mechanics of motion of a material mass is possible in this field.

3. Spaces, vectorial and tensorial fields required for the mechanics in a field. For a precise characterization of the fields which determine the motion, other spaces also are necessary.

The fibred space E defined as follows

$$E = \{R_{EM}(P) \mid P \in R\},$$

where R is the base and $R_{EM}(P)$ is the fibre corresponding to the generic point P of the base and all the spaces $R_{EM}(P)$ are isomorphic to the metric vectorial space R_{EM} may be considered as the seat of fields of a great generality.

If we choose in $R_{EM}(P)$ the vector A(P) of components $A_j(P)$ $(j = 1, 2, 3, 0)$, such that the respective potential form

$$\omega_\delta^{(p)} = \sum_{j=1,2,3} A_j \, \delta x_j - A_0 \, \delta t$$

be the scalar product in the space R_{EM}, then the field will be called a Maxwell – or classical electromagnetic – field for which we already have effected the calculations and written the equations of motion.

It follows then from the foregoing considerations that the geometrical derivative of $\omega_\delta^{(i)}$, too, will be an invariant form under the same group of transformations, hence under the Lorentz-Poincaré's group.

4. The space P of the tensor relative to a bilinear form. Let $A(A_{jk}(P);$ $j,k = 1, 2, 3, 0)$ be a matrix with 16 components, functions of the point P in the space R, and let

$$\psi(X, Y) = \sum_{j,k}^{1,2,3,0} A'_{jk} X_k Y_j - \sum_{j,k}^{1,2,3,0} A'_{jk} X_j Y_k$$

be a bilinear form, where

$$A'_{jk} = A_{jk} ; \quad j = 1,2,3; \quad k = 1,2,3$$

$$A'_{ok} = - A_{ok} ; \quad k = 1,2,3,0$$

whence

$$\psi(X,Y) = (A_{32} - A_{23})(X_2 Y_3 - X_3 Y_2) + (A_{13} - A_{31})(X_3 Y_1 - X_1 Y_3) + \\ + (A_{21} - A_{12})(X_1 Y_2 - X_2 Y_1) - \sum_j^{1,2,3} (A_{jo} + A_{oj})(X_j Y_o - X_o Y_j) .$$

5. Space of matrices. We shall denote by P the space of the matrices A which is invariant under the transformations that leave invariant the form $\psi(X,Y)$ when X and Y undergo transformations correlative to a Lorentz-Poincaré transformation of the space R_{EM}.

In particular the matrices $A_{jk} = \partial A_j / \partial x_k$, where A_j $(j = 1,2,3,0)$ are the components of a vector of R_{EM} belong to P. We put

$$\tilde{P} = \{P(\mathbf{P}) \mid \mathbf{P} \in R\},$$

where P(P) is the space P of the matrices $\omega(\mathbf{P})$ associated with the generic point P in R.

6. A tensor field. A tensor field T will be defined in P by choosing a tensor T(P) associated with each point **P** of the space R.

In particular we denote by T_M the field of the tensors

$$A_{jk} = \frac{\partial A_j}{\partial x_k} \quad (j,\ k = 1,2,3,0) \quad \text{if} \quad (A_1(\mathbf{P}),\ A_2(\mathbf{P}),\ A_3(\mathbf{P}),\ A_0(\mathbf{P}))$$

belong to the vectorial field of the space R_{EM}.

Now we can state the principles from which we derive the law of the motion of a material mass in a field.

7. The motion operator. This law must define the motion operator d.

We have in the first place a principle which operates under conditions which will be called classical.

This restricted principle has already been stated.

Let $A(\mathbf{A},A_0)$ be a vectorial field in the Einstein-Minkowski space R_{EM}, and let $\omega_\delta^{(p)} = \mathbf{A}\,\delta\mathbf{r} - A_0\,\delta t$ be the invariant scalar product, specific of this space; then the law of motion is defined by the equality

$$D\omega_\delta^{(i)} = D\omega_\delta^{(p)}$$

for an arbitrary δ.

But the right member of (3) may be written after effecting the calculation as a bilinear form

$$\sum_{j \neq k} (A_{jk} - A_{kj})(dx_k\,\delta x_j - dx_j\,\delta x_k),$$

where $A_{jk} = \delta A_j / \partial x_k$.

8. General principle. If instead of the particular Maxwell-field we consider a field T belonging to P, then the same expressions (4), where A_{jk} are only the components of the field T, will give the law of motion in a more general form

$$\frac{dp_1}{dt} = (A_{13} - A_{31})\dot{x}_2 - (A_{21} - A_{12})\dot{x}_3 + A_{01} + A_{10},$$

$$\frac{dp_2}{dt} = (A_{21} - A_{12})\dot{x}_3 - (A_{32} - A_{23})\dot{x}_1 + A_{02} + A_{20},$$

$$\frac{dp_3}{dt} = (A_{32} - A_{23})\dot{x}_1 - (A_{13} - A_{31})\dot{x}_2 + A_{03} + A_{30}$$

and the equation of energy becomes

$$\frac{d(mc^2)}{dt} = \sum_j (A_{jo} - A_{oj})\dot{x}_j .$$

Thus we obtain the law of motion in a tensor field *T*. Mesonic fields are of this type.

COMMENTARY 2

FIBRED SPACES AND THE REPRESENTATION OF MOTION

1. Generalities

The variety of the spaces which serve to describe the behaviour of physical systems in motion may cause some perplexity to the scientist who wants a clear intuition of the spatio-temporal position of the system and in particular of the material particle whose motion we have studied.

One of the fundamental principles of our investigation is identical to the Newton principle concerning the Euclidean structure of the space and time wherein we locate physical objects.

It is on that base that the laws of motion were obtained in the foregoing paragraphs; the Lorentz-Poincaré group and the four-dimensional space of Einstein-Minkowski were introduced on that base too. Do these results contradict the hypotheses?

In order to clarify the apparent difficulty which has resulted and generally to elucidate the position of the various spaces mentioned in the previous paragraph, we are compelled to introduce the concept of fibred space, the base of which remains the universe of Newton, the Cartesian product of the three-dimensional space S and the one-dimensional space T of time, and to associate with every one of its points the Einstein-Minkowski universe, at least in certain cases.

In the first part of this paragraph we shall give a general description of the fibred space and in the second part we shall deal with the fibred space of the theory of inertial motion of a material particle or, conceding to a current expression, of the theory of relativity.

2. The Universe of Mechanics as a Fibred Space

We note this space by Ω. Its fibre is the Einstein-Minkowski universe whose metrics is $d\sigma^2 = c^2\,dt^2 - dx_1^2 - dx_2^2 - dx_3^2$. Its group is the Lorentz-Poincaré group, denoted by L, and playing a twofold part:

a) it leaves invariant each separate fibre and the equations of the phenomena measured in it;

b) it represents a one-to-one correspondence between the $R_{EM}(P)$ and $R_{EM}(P')$ fibres, where $P' \neq P$, thus allowing to separate and to follow phenomena localized on different fibres which correspond to different points of the base.

The twofold function performed by the group L offers the possibility of averting different, less clear interpretations of the concept of observer and space-time reference frames bound to the observer and together with the letter to the same fibre, or better still, the concept of observers bound to different fibres.

The adoption of the notion of fibred space no longer sets the physicist in the hardly acceptable position of renouncing the intuitive initial notions of space and time bound to the Newtonian base of the fibred space.

3. The Base of the Fibred Space

This base is represented in the vast majority of physical experiments and at any scale, between the nuclear and the extragallactic, by the four-dimensional universe $S_N \times T = R$.

It is clear however, that the adoption of R as a base of the fibred space must simùltaneously satisfy two equally compelling requirements:

a) the conditions imposed by the experiment;

b) the conditions imposed by the theories adopted.

We leave aside gravity and its problems which appear, as we shall see, when we have to deal with several material masses; we shall consider only fields of the type studied before. The previously developed theory as well as the experiment, which in principle is localized in fibres, requires that the fibred space should have the structure R_{EM} and the group L.

Hence we have to investigate which are the basic spaces of the fibred space consistent with the fibre R_{EM} and whose group is L.

The result of this investigation, effected by Ion Bucur, is a topological characterization of the base B, consistent with several models, in the first place with that of Galilei-Newton; therefore we can retain the latter, at least as long as experiments do not impose another choice.

We must make a clear distinction between the law of the physical phenomenon in the course of its development and the transformations of the group, either within each separate fibre or in the correspondence between one fibre and another.

4. The Fibre

The law follows the phenomenon in its development within the base and is represented locally, for example by Maxwell's equations. But in the neighborhood of any position-moment the checking and measurement operations pertain to the respective fibre.

In this way if we follow the phenomenology in the base, we find the trace of the sequence of points of the respective fibres.

The passage from one position-moment to another position-moment realized by the law of the phenomenon may be reproduced through an L transformation between the respective fibres.

Hence, the representative space of the phenomena has eight dimensions. Its construction is based on the notion of infinitesimal connection. At the same time we associate with the fibre a differential form whose values in the Lie algebra of the group emphasize the characteristic invariants of the fibred structure and ultimately of the base in which we are particularly interested.

5. Construction of the Generators of the Characteristic Ring of the Base (after Ion Bucur)

Let L be the Lorentz-Poincaré group of rank 4, and (E, p, B, L) a principal and differentiable fibred space.

This means that all mappings are differentiable of a sufficiently high class. We want to find the differential expressions of the generators in the characteristic ring of the base. No assumption is made concerning the dimension of the variety. It may be higher than 4. The notions used for defining the Ehresmann-Lichnerowicz infinitesimal connection are the following:

T is the space tangent to the variety E at the point z; $\mathsf{L}(L)$ is the Lie algebra of the group LP ; the mapping (x, y) → [x, y] represents the function which defines the Lie algebra structure of the group ; τ_g is the homeomorphism induced by the element g of L ; τ_g leaves invariant the fibres of Ω and is induced by the translations on the right of the group L ; V_z is the subspace of T tangent to the fibre which contains z.

The family of subspaces V_z is invariant under the homeomorphisms τ_g Any vector of V_z defines a well defined element in $\mathsf{L}(L)$. If V is a vectorial space and φ : $V \to \mathsf{L}(L)$ is a linear mapping, then $[\varphi, \varphi] : V \times V \to \mathsf{L}(L)$ represents the antisymmetric linear form defined by the equality

$$[\varphi, \varphi][\xi, \eta] = [\varphi(\xi), \varphi(\eta)]$$

A connection on the fibred space {E, p, B, L } is defined as a decomposition of the space T_z, for any $z \in E$, in the form

$$T_z = V_z \otimes H_z$$

for which the following conditions must be satisfied:

a) the subspace H_z depends, in the differential form, on the point z;

b) the family of subspaces H_z is invariant under the homeomorphisms τ_g.

The elements of H_z are horizontal, those of V_z are vertical.

The form of the curvature of the connection is obtained by using the projection on the subspace V_z, given by the decomposition (1). We obtain, together with the connection $\{\Omega, p, B, L\}$, a differentiable form ω on L with values in the Lie algebra $\mathsf{L}(L)$.

Let $\alpha \in I_z$, then by definition $\omega(\alpha)$ is the element of the $\mathsf{L}(L)$ generated by the position of α in V_z.

The form of the curvature of such a connection is given by the expression

$$\Omega = d\omega + \frac{1}{2}[\omega, \omega],$$

L being a linear group, Ω emphasizes the components Ω_{ij} (i,j = 1, 2, 3, 4) which in the differential form are real and satisfy the relations

$$\Omega_{ii} = 0, \quad \Omega_{oi} = \Omega_{i4} \qquad (i = 1, 2, 3, 4)$$

$$\Omega_{ij} + \Omega_{ji} = 0 \qquad (i,j = 1, 2, 3).$$

The differential expression of the characteristic invariants results from the expansion in powers of λ of the determinant

$$\begin{vmatrix} \lambda + \Omega_{11} & \Omega_{12} & \Omega_{13} & \Omega_{14} \\ -\Omega_{12} & \lambda + \Omega_{22} & \Omega_{23} & \Omega_{24} \\ -\Omega_{13} & -\Omega_{23} & \lambda + \Omega_{33} & \Omega_{34} \\ -\Omega_{14} & -\Omega_{24} & -\Omega_{34} & \lambda + \Omega_{44} \end{vmatrix}.$$

The expansion is given by the polynomial

$$\lambda^4 + W_2 \lambda^2 + W_4,$$

where the coefficients W_2 and W_4 are of the form defined on B and generate the characteristic ring

$$W_1 = W_3 = 0,$$

$$W_2 = \Omega_{12}^2 + \Omega_{13}^2 - \Omega_{14}^2 - \Omega_{24}^2 - \Omega_{34}^2 ,$$

$$W_4 = -(\Omega_{12}\,\Omega_{34} - \Omega_{13}\,\Omega_{24} + \Omega_{14}\,\Omega_{23})^2$$

6. Final Considerations

1°.On may consider the co-chains defined by the forms W_i through integration on the chains of the space B. We obtain co-cycles, hence well determined elements in the rings of cohomology H(B,R).

The subrings thus obtained give the measure of the distance between the fibred space considered and the trivial fibred space.

2°.Newton's conceptions about time and spaces correspond sufficiently well to our conception of a fibred space.

In fact, summing up, our conception of a fibred universe consists in a base which is the S × T universe of Newton and where at each moment and position we have a universe in the sense of Einstein which is the seat and at the same time the result of our various measurements.

In support of these assertions we record Newton's reflections in the "Scolia" which follows the first eight Definitions given in "Principia".

COMMENTARY 3

FORCES AND CONSTRAINTS

1. Generalities

The second law of Newton expressed by the equality $m\mathbf{a} = \mathbf{F}$, as well as by the definitions given by him for the force of inertia, resistance, forces of acceleration or motion, forces of attraction, of repulsion, refer to what are generally called mass forces.

Other types of forces such as the force of cohesion, stresses or pressures, are not forces in the proper Newtonian sense, i.e. mass forces, and they appear as such only through interpretation and computational conventions. For example, if in classical mechanics a particle of mass m is acted on by the force $\mathbf{F}(x_1\ x_2, x_3)$ but is constrained to remain on a surface whose equation is $\varphi\ (x_1, x_2, x_3) = 0$, then we have on one hand the equality

$$(m\mathbf{a} - \mathbf{F})\delta\mathbf{x} = 0$$

and on the other hand the condition

$$\text{grad}\ \varphi . \delta\mathbf{x} = 0\ ,$$

whence the relation

(1) $$m\mathbf{a} = \mathbf{F} + \lambda\ \text{grad}\ \varphi\ ,$$

where λ and a are determined by taking into account also the equation $\varphi = 0$. The relation (1) where λ is assumed to be determined, shows that λ grad φ must be considered as representing a force: a conventional force.

By its very nature, the theory developed in the invariantive mechanics comes closer to experience and follows step by step its indications. The theory considers directly the forces whose expression is of the form $\mathbf{F} = \mathbf{E} + \mathbf{v}/c \times \mathbf{H}$ and which is specific to the forces in an electromagnetic field.

If the field H vanishes, which is the case when $\mathbf{A} = \text{grad}\ \phi$, then

$$F = \text{grad } A_o - \frac{\partial A}{\partial t} ; \qquad (2)$$

we are in the case of the electric force which reduces to a gradient when **A** is independent of t. Therefore there exists a certain generality in the expression of the force which appears in the theory expounded and it should not be necessarily interpreted as an electric or electromagnetic force. In the more general case of a non-holonomic field, the expression of the force is given in a tensorial form by equations (2); as regards the dependence on the position of the point, this is of the form

$$\frac{dp_j}{dt} = A_{jo}, \quad (j = 1, 2, 3)$$

assuming

$$A_{13} = A_{31}, A_{21} = A_{12}, A_{23} = A_{32}, A_{oj} = 0 \ (j = 1, 2, 3) .$$

The holonomic expression $\mathbf{E} + v/c \times \mathbf{H}$ or the non-holonomic one given by (2) are the only ones which represent, in the theory presented, a proper force. We further remark, that both the holonomic expression and the expression (2) may include the derivatives of the position vector, but only at the first degree. Hence, they do not correspond to the great variety of experimental problems which involve degrees higher than one, as in the case of resistance of the medium, which is proportional to the square of the velocity.

In that case we have two possibilities:

1°. We consider that $\tilde{A}(A_1, A_2, A_3, A_0)$ depends not only on the position x_1, x_2, x_3, t, but also on the velocity $(\dot{x}_1, \dot{x}_2, \dot{x}_3)$.

In that case the external derivative of $\omega_\delta^{(p)}$ has the following form

$$D\omega_\delta^{(p)} = d\mathbf{A}.\delta\mathbf{x} - dA_o\,\delta t - d\mathbf{x}.(\underset{x}{\text{grad}}\ \mathbf{A}.\delta\mathbf{x} + \underset{p}{\text{grad}}\ \mathbf{A}.\delta\mathbf{p} + \frac{\partial \mathbf{A}}{\partial t}\,\delta t) +$$

$$+ dt(\underset{x}{\text{grad}}\ A_o.\delta\mathbf{x} + \underset{p}{\text{grad}}\ A_o.\delta\mathbf{p} + \frac{\partial A_o}{\partial t}\,\delta t) .$$

Then the equality

$$D\omega_\delta^{(i)} = D\omega_\delta^{(p)}$$

gives the equations

$$\frac{dp}{dt} = \frac{dA}{dt} - \underset{x}{\text{grad}}\ A\ \frac{dx}{dt} + \text{grad}\ A_o = \underset{p}{\text{grad}}\ A\ \frac{dp}{dt} + \underset{x}{\text{grad}}\ A_o\ ,$$

(3) $$\underset{p}{\text{grad}}\ A\ \frac{dx}{dt} = \underset{p}{\text{grad}}\ A_o\ ,$$

$$\frac{d(mc^2)}{dt} = \frac{dA_o}{dt} + \frac{\partial A_o}{\partial t} \cdot \frac{dx}{dt} - \frac{\partial A_o}{\partial t}\ .$$

The last two equations must be identically satisfied on the ground of the first one, taking always into account that we have $\mathbf{p} = m\,\mathbf{x}$.

2 . The second possibility whereby forces are introduced is that of the constraints which, for generality, we consider to be of the form

(4) $$\mathbf{C}.\delta\mathbf{x} + \mathbf{D}\delta\mathbf{p} + L\,\delta t = 0\ ,$$

where C, D and L depend on x, p and t. One may have several constraints of the preceding type.

Taking then into account the principle of motion

$$D\omega_\delta^{(i)} = D\omega_\delta^{(p)}$$

and condition (4), the resulting equations are

$$\frac{dp}{dt} = \underset{p}{\text{grad}}\ A\,\frac{dp}{dt} + \underset{x}{\text{grad}}\ A_o + \lambda C\ ,$$

$$\underset{p}{\text{grad}}\ A\ \frac{dx}{dt} = \underset{p}{\text{grad}}\ A_o + \lambda D\ ,$$

$$\frac{d(mc^2)}{dt} = \frac{dA_o}{dt} - \frac{\partial A_o}{\partial t} + \frac{\partial A}{\partial t}\,\frac{dx}{dt} - \lambda L\ ,$$

to which we must add, for the determination of λ the relation obtained from (4), where we take d instead of δ,

$$\mathbf{C}\,\frac{dx}{dt} + \mathbf{D}\,\frac{dp}{dt} + L = 0\ .$$

2. Force and Gravity

Gravitation is the current expression for the force corresponding to the interaction resulting from the presence of several material masses. A gravity field is the result of the presence of a mass in the neighbourhood of other masses. This field and the resulting force have no correspondent in the preceding theory. They are manifest in Einstein's theory in a geometrical form. The presence of a mass modifies the structure of the space, gives it a special curvature, modifying thus the laws of motion.

In the theory presented however it is not the basic space that has its structure modified, but the inertial elements of the bodies in presence.

Obviously, we can always point out with a certain approximation a field and a force, useful in calculations, and which we may interpret as a gravity field or force.

NOTE
MOTION OF THE PERIHELION

Ieronim Mihaila

From the formulae of chap. III, neglecting the interaction mass ν , we obtain the equations of motion of the sun and planet, namely

$$\frac{dp_1}{dt} = \mathbf{L}, \quad \frac{dp_2}{dt} = - \mathbf{L}, \tag{1}$$

where

$$\mathbf{p}_1 = m_1 \mathbf{v}_1 + \mu \mathbf{v}_2 , \ \mathbf{p}_2 = m_2 \mathbf{v}_2 + \mu \mathbf{v}_1 , \tag{2}$$

$$\mathbf{L} = 2 c^2 (1 - \mathbf{v}_1 \mathbf{v}_2 / c^2) \frac{\mathbf{r}}{r} \frac{\partial \mu}{\partial r} . \tag{3}$$

In order to obtain the equations of relative motion of the planet with respect to the sun, we seek a preferential inertial reference frame. To this let us consider the point whose position vector is ρ,

$$\rho = \frac{(m_1 + \mu)\mathbf{x}_1 + (m_2 + \mu)\mathbf{x}_2}{m_1 + m_2 + 2\mu} . \tag{4}$$

Since $H = c^2 (m_1 + m_2 + 2\mu) = \text{const.}$, we obtain

$$\overset{\circ}{\rho} = \frac{\mathbf{p}_1 + \mathbf{p}_2}{m_1 + m_2 + 2\mu} + \frac{\dot{m}_2 + \dot{\mu}}{m_1 + m_2 + 2\mu} \mathbf{r} . \tag{5}$$

We shall now show that $\rho = \text{const.}$, if we confine ourselves to the terms in v_1^2/c^2 and v_2^2/c^2.

Indeed, equations (1) become

$$\frac{d}{dt} [m_1^0 \dot{\mathbf{x}}_1 (1 + \frac{1}{2} \frac{\dot{\mathbf{x}}_1^2}{c^2}) + \mu_0 \dot{\mathbf{x}}_2] = \frac{f m_1^0 m_1^0}{r^2} \frac{\mathbf{r}}{r} (1 - \frac{1}{2} \frac{\dot{\mathbf{x}}_1 \dot{\mathbf{x}}_2}{2c^2})(1 + \frac{2k}{c^2 r}),$$

$$\frac{d}{dt} [m_2^0 \overset{\circ}{x}_2 (1 + \frac{1}{2} \frac{\dot{\mathbf{x}}_2^2}{c^2}) + \mu_0 \dot{\mathbf{x}}_1] = - \frac{f m_1^0 m_2^0}{r^2} \frac{\mathbf{r}}{r} (1 - \frac{1}{2} \frac{\dot{\mathbf{x}}_1 \overset{\circ}{\mathbf{x}}_2}{\epsilon c^2})(1 + \frac{2k}{c^2 r}), \tag{6}$$

where

$$\mu_o = - \frac{1}{2} \frac{f m_1^0 m_2^0}{c^2 r}$$

and $\lambda(r)$ is taken of the form k/r, k being a constant (cf. I. Mihaila, C.R. Acad. Sc. Paris, A, 280, 595-598, 1975).

We obtain

$$\dot{\rho} = \frac{p_1 + p_2}{m_1 + m_2 + 2\mu_o} + \frac{m_2^0}{m_1 + m_2 + 2\mu_o}\left(\frac{v_2 \cdot \dot{v}_2}{c^2} + \frac{1}{2}\frac{fm_1^0 \dot{r}}{c^2 r^2}\right) r$$

$$= \frac{p_1 + p_2}{m_1 + m_2 + 2\mu_o} + \frac{m_2^0}{m_1 + m_2 + 2\mu_o}\left(-\frac{fm_1}{c^2 r} v_2 + \frac{1}{2}\frac{fm_1^0}{c^2 r}\frac{\mathring{r}}{r} r\right)$$

i.e., since the relative motion is nearly circular,

$$\dot{\rho} = \frac{p_1 + p_2}{m_1 + m_2 + 2\mu_o} - \frac{m_2^0}{m_1 + m_2 + 2\mu_o} v_2 \frac{\dot{r}^2}{c^2} + \frac{1}{2}\frac{m_2^0}{m_1 + m_2^0 + 2\mu_o} \dot{r} \frac{\dot{r}^2}{c^2}\frac{r}{r}.$$

For most planets we have $m_2^0 / m_1^0 < |\dot{r}| / c^2$, and therefore we may consider

$$\dot{\rho} = \frac{p_1 + p_2}{m_1 + m_2 + 2\mu_o} = \text{const.} \tag{7}$$

In other words the reference frame with the origin at the point of vector ρ is, within the approximation considered, an inertial frame.

In this reference frame we have the relations

$$\begin{aligned} (m_1 + \mu_o) r_1 &= -(m_2 + \mu_o) r_2 , \\ (m_1 + \mu_o) \dot{r}_1 &= -(m_2 + \mu_o) \dot{r}_2 , \\ (m_1 + \mu_o) r_1 &= (m_2 + \mu_o) r_2 , \end{aligned} \tag{8}$$

where r_1, r_2 are the position vectors, and the equations of motion take the form

$$\begin{aligned} \frac{d}{dt}\left[\dot{r}_1 \left(1 + \frac{1}{2}\frac{\dot{r}_1^2}{c^2} + \frac{1}{2}\frac{fm_1^0}{c^2 r}\right)\right] &= \frac{fm_2^0}{r^2}\frac{r}{r}\left(1 + \frac{1}{2}\frac{m_1^0}{m_2^0}\frac{\dot{r}_1^2}{c^2}\right)\left(1 + \frac{2k}{c^2 r}\right), \\ \frac{d}{dt}\left[\dot{r}_2 \left(1 + \frac{1}{2}\frac{\mathring{r}_2^2}{c^2} + \frac{1}{2}\frac{fm_1^0}{c^2 r}\right)\right] &= -\frac{fm_1^0}{r^2}\frac{r}{r}\left(1 + \frac{1}{2}\frac{m_2^0}{m_1^0}\frac{\mathring{r}_2^2}{c^2}\right)\left(1 + \frac{2k}{c^2 r}\right). \end{aligned} \tag{9}$$

Keeping only the second order terms in $|r|/c$ and subtracting equations (9), we obtain

$$\frac{d}{dt}\left[\dot{r}_2 \left(1 + \frac{1}{2}\frac{\mathring{r}_2^2}{c^2} + \frac{1}{2}\frac{fm_2^0}{\epsilon c^2 r}\right) - \dot{r}_1\right] = -\frac{f(m_1^0 + m_2^0)}{r^2}\frac{r}{r}\left(1 + \frac{2k}{c^2 r}\right). \tag{10}$$

Using the relations

$$\dot{\mathbf{r}} = \dot{\mathbf{r}}_2 - \dot{\mathbf{r}}_1 \ , \quad \dot{\mathbf{r}}_2 = \frac{m_1 + \mu_o}{m_1 + m_2 + 2\mu_o} \dot{\mathbf{r}} \ ,$$

we may write

$$\dot{\mathbf{r}}_2 \frac{\dot{\mathbf{r}}_2^2}{c^2} = \dot{\mathbf{r}} \frac{\dot{\mathbf{r}}^2}{c^2} \left(\frac{m_1}{m_1 + m_2 + 2\mu_o} \right)^3 = \dot{\mathbf{r}} \frac{\dot{\mathbf{r}}^2}{c^2} \ .$$

Equation (10) becomes

$$\frac{d}{dt} \left[\dot{\mathbf{r}} \left(1 + \frac{1}{2} \frac{\dot{\mathbf{r}}^2}{c^2}\right) \right] = - \frac{f(m_1^0 + m_2^0)}{r^2} \frac{\mathbf{r}}{r} \left(1 + \frac{2k}{c^2 r}\right).$$

We obtain

$$\text{(11)} \qquad \ddot{\mathbf{r}} = - \frac{f(m_1^0 + m_2^0)}{r^2} \frac{\mathbf{r}}{r} \left(1 - \frac{3}{2} \frac{\dot{\mathbf{r}}^2}{c^2}\right)\left(1 + \frac{2k}{c^2 r}\right)$$

or .

$$\text{(12)} \qquad \ddot{\mathbf{r}} = - \frac{\mu^*}{r^2} \frac{\mathbf{r}}{r} \left(1 - \frac{2\alpha\mu^*}{c^2 r} + \frac{2\beta\dot{\mathbf{r}}^2}{c^2}\right),$$

where

$$\text{(13)} \qquad \mu^* = f(m_1^0 + m_2^0) \ , \quad 2\alpha\mu^* = -2k, \ 2\beta = -3/2 \ .$$

One sees that the motion is plane. Taking the plane of the motion as a reference plane, the equations of motion become

$$\text{(14)} \qquad \ddot{x} = - \frac{\mu^* x}{r^3} + X, \quad \ddot{y} = - \frac{\mu^* y}{r^3} + Y,$$

where

$$\text{(15)} \qquad X = \frac{\mu^* x}{r^3} \left(\frac{2\alpha\mu^*}{c^2 r} - \frac{2\beta \dot{r}^2}{c^2}\right),$$

$$Y = \frac{\mu^* y}{r^3}\left(\frac{2\alpha\mu^*}{c^2 r} - \frac{2\beta r^2}{c^2}\right). \tag{16}$$

Because the force (X, Y) is small, we obtain as a first approximation the equations of elliptical motion. By integration we obtain the osculating orbit. The action of the additional (corrective) force may be considered as a perturbation and the motion of the perihelion may be studied by the method of the variation of constants (e.g. see J. Chazy, La théorie de la Relativité et la Mécanique céleste, t.I, chap. [I], Gauthier-Villars, Paris, 1928).

The differential equation of the longitude of the perihelion is of the form

$$\frac{dw}{du} = \frac{\sqrt{1-e^2}\,(1-e\cos u)}{n^2 a^2 e}\left\{X\frac{\partial x}{\partial e} + Y\frac{\partial y}{\partial e}\right\}, \tag{17}$$

where a is the semi-major axis of the osculating orbit, e the eccentricity, n the mean motion (mean angular velocity) and u the eccentric anomaly.

The derivatives $\dot{x}$ and $\dot{y}$ are computed by the formulas of elliptic motion, where the time and the eccentric anomaly are considered as variables. The derivatives of the coordinates with respect to e are computed by the same formulas considering the eccentric anomaly and e as variables. In the plane of the motion the formulas of elliptical motion are

$$\begin{aligned} x &= a(\cos u - e)\cos w - a\sqrt{1-e^2}\,\sin u\,\sin w,\\ y &= a(\cos u - e)\sin w + a\sqrt{1-e^2}\,\sin u\,\cos w,\\ r &= a(1 - e\cos u),\\ u - e\,\sin u &= nt + \ell_0 - w, \end{aligned} \tag{18}$$

where ℓ_0 is the mean longitude at t = 0.

The equation of the longitude of the perihelion becomes

$$\frac{dw}{du} = \frac{\mu^*\sqrt{1-e^2}}{c^2 a e^2}\left\{\frac{2\beta - Y\lambda}{1-e\cos u} + \frac{2\alpha - 4\beta - 2\beta(1-e^2)}{(1-e\cos u)^2} + \frac{(-2\alpha + 4\beta(1-e^2))}{(1-e\cos u)^3}\right\}. \tag{19}$$

The increment corresponding to the time interval wherein the eccentric anomaly varies from u to u + 2π is obtained by integrating (18) between these limits. We obtain

$$\delta w = \frac{2\pi\mu^*}{c^2\, a(1-e^2)}\,(-\alpha + 2\beta)\,. \tag{20}$$

In the solar system, after the elimination of planetary perturbations, the perihelion shows a direct motion which is well represented by the relation

$$\delta w = \frac{6\pi\mu^*}{c^2\, a(1-e^2)}\,. \tag{21}$$

Comparing (20) and (21) we obtain the condition

$$-\alpha + 2\beta = 3\,. \tag{22}$$

Introducing the values (13), we obtain

$$k = \frac{9}{2}\mu^* = \frac{9}{2}\, f(m_1^0 + m_2^0)\,. \tag{23}$$

and consequently

$$\varphi(r) = -\frac{f m_1^0 m_2^0}{2c^2 r}\left(1 + \frac{\lambda(r)}{c^2}\right), \tag{24}$$

where

$$\lambda(r) = \frac{9}{2}\, f\,\frac{m_1^0 + m_2^0}{r}\,. \tag{25}$$

and the corrective term of Newton's law is determined.

CONTENTS

Printed in the United States
By Bookmasters